I0833135

NEVER VANQUISHED

NEVER VANQUISHED

A NOVEL OF THE VIETNAM WAR

DOUGLAS YOUNG

Deeds Publishing | Athens, GA

Published by Deeds Publishing in Athens, GA
www.deedspublishing.com

Printed in The United States of America

Cover photo by Robert O. Babcock, October 1966, Montagnard village, east of Pleiku

ISBN 978-1-961505-58-2

Books are available in quantity for promotional or premium use. For information, email info@deedspublishing.com.

First Edition, 2026

10 9 8 7 6 5 4 3 2 1

Dedicated to My Family

For Cindy, the wife I met in Vietnam

For Keith, the son born while I was in Vietnam

For Trang, daughter of Vietnam

For Ai Nhan, daughter of Vietnam

PREFACE

This book is wrapped around history, but the entire story is certainly not factual. A large portion of it is documented history. For the most part, it's up to the reader to figure which is which. Here's some help.

The character of DeAndre Washington is fictitious. FULRO was a real entity. The Jarai are one tribe of the mountain people in Vietnam referred to as "Montagnards." They thrive today, both in their native lands and in North Carolina. The character of Taylor "Smithy" Smith is based on an actual Christian missionary to the Jarai. The date of the transfer of the Pleiku Air Base is depicted at a different time than it actually happened. I have blended two South Vietnamese generals into one person: Gen. Vinh Loc who commanded II Corps from 1965 to 1968, and General Ngô Du, who commanded II Corps from August 1970 onward. Both despised and persecuted the Montagnards.

If the reader wishes to know more about those times, I encourage them to research it. I have included a bibliography and some Internet links at the end.

CHAPTER ONE

"The War" loomed close during the Sunday worship service at the small white church on 26th Street. The clapboard building had stood there for generations, its sanctuary infused with the scents of polished wood and snuffed-out candles. A stained-glass cross inside a window stood above the pulpit. Like every Sunday, it was full.

Every person in the building that morning—everyone—laid hands on DeAndre as Pastor Willis fervently pleaded for God's protection during yet another tour of duty in Vietnam. His mother stood beside him, tears spilling unchecked down her cheeks, her grip on his hand a desperate tether to the earth. She didn't sob; her grief was deeper than that. "Father, we implore you to protect this young man as he serves his country, that he returns to his family to once again worship you in this church." The congregation erupted in shouts of "Praise God!" and "Amen, Pastor!"

The Washington family had belonged to this little congregation for generations. As the service concluded, Pastor Willis greeted congregants at the front door, accepting their usual compliments on his sermon. Outside, the sun beat down, and the other churchgoers edged away from DeAndre, granting him a semblance of privacy as he attempted to console his mother. Fear, sadness, disbelief, and fury all showed in her tear-streaked eyes.

"But we're pulling out of there. Why are they sending you now?" she sobbed, her voice trembling. She had seen the stories on television—wives running across the tarmac to hug their husbands, new fathers meeting their children for the first time, politicians assuring the nation that the war in Vietnam would soon be a thing of the past. But her son was going back.

Mrs. Jada Washington's heart burned with resentment toward the powers that had thrust America into war, yet continued to send her son to fight. It was January 1972, and the conflict in Southeast Asia had stretched on since John Kennedy was president back in 1963.

"Dashawn only served one year over there," she continued, her voice sharp with indignation. "It isn't right that they'll send you there more than your brother."

DeAndre let the comment about his younger brother slide, the unspoken tension hanging in the air. Dashawn had served as a Platoon Leader with the 198th Infantry Brigade in central South Vietnam, facing his share of firefights but returning home unscathed, and out of the Army. "I'll go back now because we're training the Vietnamese to fight the war themselves," he said, attempting to reassure her. "I've been assigned to advise one of their battalions. I won't be fighting."

He saw the glint in her eyes that said she wasn't buying it—not for a second. "You think I don't know you?" she pressed, her voice low but fierce. "You don't push paper. You lead. You're in the thick of it. You run things, DeAndre. You don't sit back and watch."

The southern sun felt oppressively warm as DeAndre forced a smile, trying to lighten the mood. Cheerfully waving to the congregation, he pulled his mother away and into the car for the drive back to her house. Inside, she prepared his favorite lunch—fried chicken, collard greens, and sweet potatoes—a feast that felt bittersweet as it filled the air with comforting aromas.

After changing out of his Sunday-go-to-meeting clothes into a Class A uniform, he called Susan and asked her to take him to the airport. He feared that if his mother drove, her emotions would lead to an accident. One last hug, followed by a heartfelt, "I love you, Mama," and he stepped out, carrying only a small duffle.

"Thanks for coming to get me, Susan," he said as he settled into the passenger seat. "Then again, I never pass up an opportunity to see you."

"Well, I hope you got an eye full a couple of nights ago," the statuesque brunette replied with a playful grin. "It's going to be a while before you get the chance again."

The two had known each other since college, and Mrs. Washington had hoped, albeit reluctantly, that their friendship would blossom into romance. Yet, she also understood the complications of DeAndre having a white wife in a racially charged world.

As Susan's car disappeared down the street, Mrs. Washington sighed and retreated to the coolness of her living room, settling into her soft easy chair. Photos of her two sons and her late, beloved husband surrounded her, each frame a reminder of love and loss. Maurice had been the only man she ever loved, taken from her far too soon. Both her sons had served in combat—one had returned home, but the other seemed determined to emulate his father's fate and die young.

She knew that wasn't entirely true; DeAndre was a complex mix of do-gooder and disciplinarian. He held strong convictions about the South Vietnamese cause and was willing to risk his life for it. An adrenaline junkie, the Army had only intensified his thirst for thrills. He first tasted the rush of jumping from perfectly good airplanes, but the addiction deepened during his first tour in Vietnam, where he had seen combat and earned a Purple Heart.

Even though he was a good officer, he was no martinet, a bit soft-hearted. DeAndre sought respect, not fear, from his subordinates. Her complex son defied her wish for him to find a good woman, settle down, and lead a quiet life—no woman had been able to compete with the Army so far.

It wasn't a long drive to the airport, but Susan took her time. "Isn't this supposed to be some sort of non-combat assignment? I mean, are you going to be in an office in a city?"

"Not quite," he replied with a chuckle, trying to ease her worries. "I'll be advising one of the South Vietnamese units. I won't be leading them, and I'm not supposed to fight. I certainly won't be in any city."

"Okay, okay. Maybe a small city or town. I just don't want to think of you getting tight with some exotic Asian woman," she said, playfully hiking the hem of her skirt a little higher.

"Exotic? Hardly," he scoffed. "By the time Vietnamese girls are sixteen, they've been working in the rice paddies for six or seven years. By sixteen, they already have a wrinkled, weathered look about them. Appealing? I don't think so."

"Uh huh. Sure," she teased, pulling over to the curb a short distance from the terminal. He slid closer on the bench seat of her old Chevy, his left arm wrapping around her shoulder as he pulled her in, their lips meeting in a soft kiss. A moan escaped her as his hand grazed her breast, but they broke apart too soon. "Take care of yourself," she whispered, her eyes searching his for reassurance.

"I will," he promised, stepping out to retrieve his duffle from the trunk. He cast a final glance at her, their shared moment lingering like a ghost between them.

Inside the small airport, the atmosphere buzzed with the hum of activity. DeAndre quickly checked in, the process smooth and efficient. As he made his way to the gate, he noticed most Americans regarded him with respect, nodding politely. But not everyone shared that sentiment; one woman hurried past with her small child, her expression shifting to one of distrust as she clutched her son tighter, crossing to the other side of the walkway when she saw the black officer. He felt her rejection but pushed it aside, reminding himself that in combat, the only important color was olive green.

CHAPTER TWO

The chartered Boeing 707 was on the tarmac at Travis Air Force Base north of Oakland, California. An olive drab Army bus picked up passengers at the terminal, taking them to the plane. DeAndre walked down the aisle until he spotted an older Chief Warrant Officer whose uniform said he's been to Vietnam before. After asking the preliminary "This seat taken?", he sat down. The chatter quickly switched to going back.

"Yeah, this is my third tour, Chief," said DeAndre. "You?"

"Me too. I don't know what they're gonna have me do. I'm a logistics guy, and all I hear about is all the stuff being given to the South Vietnamese."

"What are the ARVN getting?" asked the captain, using the usual short-talk for the Army of the Republic of Vietnam.

"Damn - you name it. I've got a buddy assigned to a hospital. Besides the building itself, we gave them all the beds, the mess hall, the motor pool, the operating room equipment—yeah, even the vehicles."

"Now that's what I call Vietnamization. Hell, way back in '63, McNamara said that was a South Vietnamese war."

"Yep—the time has come for the South Vietnamese to fight the war themselves."

"I know I'm going to be an advisor—all the American line units are going home. That's cool. I've never worked with Marvin the ARVN before and I'm a little nervous about not being able to speak Vietnamese."

"I thought they sent all advisors to the Defense Language School at Monterey."

"Not anymore. I hope the guy I work with has been to the US for training so he can speak at least a little English."

The conversation went on for some time as they complained about the Army as only two careerists could do. After a while, the pauses between sentences got longer and eventually both drifted off into sleep.

The big airplane settled onto the runway at the American base at Cam Ranh Bay, ending the long trip across the Pacific. There'd been time to stretch his legs during the refueling stop at Guam, but DeAndre was still happy to be able to walk around again.

The plane rolled to a stop. "Hey, Chief. Thanks for the company. It was nice to be able to talk to somebody else who's been here before."

As he walked towards the terminal, he realized his mom was right—of course. On the one hand, he hated being away from his widowed mother, yet he also looked forward to the sense of adventure battle gave him. His previous combat tours had increased the compulsive desire for excitement. He enjoyed masculine company, though he wasn't sure how comfortable he'd be surrounded by people speaking a different language and eating different food.

True, he'd have two other Americans with him and there would be the Armed Forces Vietnam Radio Network to bring him news and music, but he would be pretty isolated for a year.

Early January 1972 was a good time to be in Cam Ranh. The sun was bright, but the temperature was in the low eighties. There were an equal number of Americans and Vietnamese on the big base. The Vietnamese were taking it over, both the deep-water harbor that sheltered war and cargo ships, as well as the big airfield. There were few American ground combat units left in-country. Sure, there was an understrength brigade of the 1st Cavalry Division at Bien Hoa just north of Saigon, and the 196th Light Infantry Brigade was still at Danang, plus a smattering of other units in the process of packing up to go home.

By President Nixon's edict two months ago in November, Americans had stood down from any offensive operations. There were a few more aviation units left, mostly to provide transport for American advisors and provide medical evacuation to the few remaining American hospitals. The Air Force still had tactical air support units flying out of Saigon and Danang, as well as the giant B-52s in Guam.

DeAndre was greeted in the terminal by Sergeant First Class Eric Wozniak, who informed him he was the senior NCO assigned with him to the 11th Battalion of the Vietnamese Rangers. Though the parent 2nd Ranger

Group was headquartered at Camp Holloway, near Pleiku, the 11th Battalion was an hour away at Kon Tum up in the mountains of the Central Highlands.

Wozniak told him he'd left a young sergeant back at the base camp to hold down the fort until they returned. Washington knew instantly that he and the grizzled old NCO were going to get along famously. Both were on their third deployment, both loved being in the infantry, and both were Master Parachutists. Washington had the whimsical thought that they might be the last Americans left in Vietnam, tasked with turning out the lights in Saigon.

A quick jeep ride to the MACV administrative area on the base and Washington was processed in, issued his TA-50 field gear, three camouflage uniforms the same as the Rangers wore, and an M-16, followed by a short but cordial conversation with the colonel who headed up the advisory teams in II and III Corps. Another jeep was called and the captain was taken to board a twin-engined Army Caribou airplane for the flight to Camp Holloway.

As they bounced back to the airfield, Washington and Wozniak talked about how lucky they were to be assigned to the Rangers instead of a regular South Vietnamese army unit.

Wozniak had been with the 11th for a while. "Rangers are not the "Marvin the ARVN " draftees seen in typical South Vietnamese units, sir. They are solid troops."

Captain Washington hoped the officers he would be advising were professionals and not Saigon cowboys who only wanted to return to the capital to get their share of the corruption and politics. "Oh, I don't think our advisees are like that, sir. I've seen their CO in a fight and he's pretty fierce."

The Caribou was a very noisy airplane, making conversation all but impossible. They both settled into the canvas seats along the side of the aircraft, being sure their feet weren't caught in the mish-mash of netting covering the cargo loaded in the middle. They'd have an hour in the air to Camp Holloway, followed by a much shorter trip by Huey helicopter from there to Kon Tum. It would be a long day, especially since he'd just come across the Pacific.

Yes, Captain Washington had been in Vietnam before. Back in 1969 and 1970, he'd had the privilege of commanding a company in the justly famous 1st Cavalry Division. He'd been a platoon leader in the highly decorated 173rd Airborne Brigade in 1966 and 1967. He figured a year of advising should be a cakewalk compared to his two other tours.

He'd already been told he would be going to a university to earn his master's degree when he came home, and that was a prerequisite for promotion to major. His two Silver Stars and a Purple Heart, plus copious command time in combat, coupled with "walk on water" evaluations should cement his advancement.

Maybe after this tour, he would think about finding a wife. He knew he was lonely for a woman's touch. He'd watched his parents while growing up and knew they completed each other. But finding a woman who would tolerate his military career would be tough to do. So yeah—he'd find a wife.

But not now.

Hoping to get some rest during the flight, Washington closed his eyes before the plane started to roll. His face relaxed and the smile lines around his eyes became more prominent. Women often complimented him on the rich chocolateness of his skin but kept it to themselves that they thought him ruggedly handsome.

He didn't have a particularly muscular body, but rather one that might be described as sinewy, with almost no fat on his six-foot frame. It was his demeanor that surprised people. Many assumed that anyone who had seen so much combat would be sort of a tough guy, but DeAndre was modest to the point of being slightly shy and was unfailingly courteous to all. Women adored him; men admired him.

DeAndre drifted off into an easy snooze. Sleep often brought old memories to his mind. In his dreams, he remembered being a first-generation college student. In his segregated high school, he'd coupled his innate intelligence with a strong work ethic. The result was that not only did he gain admission to a very good school at the University of Georgia, but he got a full scholarship as well.

Once he got used to the idea that he was a raisin in a bowl of Cream of Wheat, he made friends easily. Sure, he had the usual college bouts of drunkenness and girl chasing, but he kept up good grades. There weren't many black women on campus, but he and Susan were regulars. He liked ROTC and went on to become an Army officer that way.

He'd surprised his superiors by asking to be commissioned in the Infantry. Most Reserve Officers asked for assignment to the Finance Corps or Medical Service Corps or other non-combat postings, but DeAndre wanted to be at the center of things. The Infantry was for him, as was Airborne School and Ranger training.

That's when his mom began to worry about him. She was not keen on him jumping out of perfectly good airplanes, being a so-called "snake eater," or even being in the Infantry. Her birthday present to him prior to his first deployment was a book titled *"How to Survive in Vietnam."*

DeAndre snapped his neck as the Caribou bounced down the runway at Pleiku. "Damn—I've been sleeping a lot harder than just a snooze," he thought as he unfastened his seat belt; but he waited until the crew chief indicated it was okay to stand up.

Retrieving their duffel bags, he and SFC Wozniak walked down the ramp into the blazing sun of Vietnam's Central Highlands, squinting to see the jeep they'd been told would be there to pick them up.

Amid all the noise on the ramp, they heard a voice say, "Over here, Captain." The young man approached the two and threw their duffels in the back of a jeep. "Colonel Arredondo wants to see you before you go on to Kon Tum, sir."

DeAnde was not surprised that the colonel wanted to see him: the Old Man wanted to meet and assess the newbie. The jeep ride was so short that DeAndre almost wondered why the colonel had bothered to send a vehicle. Not far to the rear of Pleiku's Camp Holloway airfield's terminal was a simple sand-bagged medium tent, reinforced by a 2X4 frame. DeAndre smiled as he realized his new boss was not into enjoying all the privileges of rank he was entitled to. Leaving their duffle bags in the jeep, DeAndre flipped open the tent flap, but before his boots touched the wood floor made from shipping pallets, he was greeted with a loud, "Damned good to see you, Captain Washington."

Before he'd had a chance to salute properly, Lieutenant Colonel Rodolfo Arredondo was standing in front of him with an extended hand, ignoring DeAndre's salute. "Have a seat, Captain. Sergeant Wozniak treat you okay?"

DeAndre settled into the chair, noticing the Combat Infantryman's Badge, master blaster jump wings, and Ranger tab on the colonel's camouflage uniform. The two settled into an easy conversation that lasted much longer than the usual "Welcome, FNG" chat.

Like DeAndre, the colonel had been in Vietnam before (and, truth be known, in Laos back when the US didn't have any troops in Laos.) Unlike DeAndre, he had a lot of experience working with both the Vietnamese and the hill tribes known to Americans as Montagnards.

The two men sized each other up, both liking what they saw. To Captain Washington, the colonel seemed like a straight-up guy who wasn't a jerk when telling you what he thought and wanted. He also sensed that his boss was as emotional as algebra, which is a good thing when the shit hits the fan. To LTC Arredondo, his new officer was not a funny new guy, but competent and likable.

"I think you're going to like working with the 11th",

said the colonel. "Unlike so many other units, the officers actually give a shit and are fairly competent. They aren't Saigon cowboys."

Washington was a little surprised to hear, "Most Rangers are excellent fighters, but they don't always mind killing other Vietnamese if their leaders tell them the other guy's ethnicity is wrong." Arredondo went on to say that most officers were promoted for their loyalty to whatever group of generals were in power in Saigon, currently General Nguyen Van Thieu, rather than promoted for their competence.

"By the way, you may not know it, Captain, but the US no longer assigns advisors to regular battalions in the ARVN, but so far, they have made the Marines, the Airborne, and Rangers an exception. I expect even that will end soon, but the big shots haven't seen fit to fill me in with any details."

"Any ideas what will happen to us 'leftovers' when that happens, sir?"

"Who knows? They may just move you up to group level or make you the Assistant Officer in Charge of Paperclips at MACV headquarters. They may even send you home early. That's already happening in some line units."

The colonel went on, asking, "How much have you worked with the Vietnamese before, Washington?"

"Not a bit, Sir. I was with the 173rd on my first tour, then the 1st Cav on my second. Anytime we saw Vietnamese, we shot them."

The slightest hint of a smile tugged at the corners of Arredondo's mouth. Washington's sense of humor was a bit heavy, but not unexpected for someone who had served two combat tours as a company-grade officer in line units.

"Okay, you'll get used to it soon enough. It's just hard to remember sometimes that you don't have command authority—you are what your title says—an advisor. You'll probably like your battalion commander, Major Hoang Nien. By the way, the Vietnamese name order is the reverse of ours. Hoang is his family name, but you address him by his given name. Call him Thiếu tá Hoang, saying the rank as "tyoo tah."

"But, there are two things to watch for. First, he is very, very prejudiced against the Jarai. In fact, I'd call him brutal when it comes to the Jarai."

"Excuse me, sir, but who are the Jarai?"

"Ah, yeah. To the vast majority of Americans, the people who live here in the Highlands are Montagnards. That word comes from the French, who once colonized this country. It simply means mountain people. There are fifty some-odd different tribes, and they speak a lot of different languages and have very different customs from the Vietnamese. A lot of them are Christian. The Jarai are one of the bigger tribes and they live around here. Our Special Forces work a lot with them."

"Why does Thiếu tá Hoang hate them?"

Arredondo stood up and paced a few steps. He was a big man and obviously very fit. The first harbingers of gray were peeking out of his close-cropped hair and added to the air of ruggedness that went along with his tanned and somewhat leathery face. He smiled at Washington's question, paused, then went on.

"Think of American Indians in the 1800s. The white men looked at all that land and wanted it for themselves. The US kept making treaties, most of which were forced on the Indians, and then the government broke the treaties. White men called them savages. The Vietnamese call the Montagnards moi—which means savage. It's a common attitude. It's just that Hoang doesn't even try to disguise his feelings. The Vietnamese government encourages such attitudes among officers. The Vietnamese want the land. It's great for growing coffee and tea."

The colonel was still standing, obviously pondering what he would say next. Once he decided to say it, his confidence rang strong. "Washington, as a black man, you know your people's history. The original US Constitution said that black slaves were only three fifths of a person—in other words, not quite human. You know better than I that some white people still feel that way."

DeAndre was startled that his new boss would be so up front about a topic that most avoided as being too sensitive, but he said nothing as he could see the colonel had more to say.

Arredondo went on to explain the complex and bloody history between the Jarai and Vietnamese. "You remember President Diem of South Vietnam—the one who was overthrown and assassinated in 1963 about three weeks before President Kennedy was killed? Well, Diem was particularly brutal in dealing with the Montagnards, especially the Jarai.

"A few years ago, back in the 60s, some of the hill tribe troops rebelled against their Vietnamese officers, holding some of them captive. That gave rise to *Front uni de lutte des races opprimées*, abbreviated FULRO. When the French ran the country, the 'Yards pretty much had control of their own fortunes, but when the Vietnamese took over, they called them 'moi', wouldn't allow tribal languages to be taught in schools, and Vietnamese troops openly stole stuff from people's houses."

DeAndre pondered the treatment of the Jarai by the Vietnamese. He could identify. As a black man, he'd been treated badly a number of times.

"But, there is one more thing I want you to be careful of. While I really think Hoang is competent, I also think he is crooked. I'm not sure how he is doing it, but he's making a lot of money that's above his salary. Just keep an eye open for that. And be sure you never mention FULRO around him, unless you want to see him go ballistic. FULRO still exists. The men are well trained and armed by our Special Forces. They want an independent country for the mountain tribes."

With that, SFC Wozniak was called in and the three bantered a bit. "Okay, Sergeant—you take good care of our tenderfoot captain here. He's never worked with Vietnamese before, so he'll need to be broken in properly."

The three chatted a while longer, until the colonel signaled it was over by saying, "Washington, the next time you come to beautiful downtown Pleiku, we'll sip a little of the good Scotch I just happen to have. I hope you're a Scotch drinker. I know Sergeant Wozniak likes it—I've seen him gulp down half of my stash."

Some manly back slapping and guffaws ensued, and

it was obvious to Washington that he was part of a solid team.

DeAndre and Wozniak waited for the jeep to take them back to the helicopter pad. Though they both thought the distance was too short to need a ride, their gear was in the jeep. The helicopter crew chief and jeep driver helped toss their bags into the bird, then the engine whined and the blades began to turn slowly. They waited until the crew had finished their checklist, then the pilot pulled pitch and headed towards Kon Tum.

The Huey helicopter was notoriously noisy—much too noisy to talk—so DeAndre settled for watching the landscape beneath him. He'd never seen Vietnam like this. Here and there were little pockets of puffy whiteness, the remnants of morning mist. The jungle itself was verdant, so brilliantly green that it almost glowed. He saw little else during the hour-long trip, except for very short glimpses of what he assumed were villages.

The buildings, as much as he could tell, had thatched roofs and were very long. But it was the mountains that surprised him. These were not just oversized hills like he'd seen west of Saigon. The highest one he saw seemed to be somewhere north of KonTum. His map told him it was Ngoc Linh and it was 8,524 feet high.

CHAPTER THREE

Soon, the bird made a big wide turn, passing over a looping river, as it descended over Kon Tum. DeAndre saw a military compound on the north side of the small city, with what must be Route 14 on the east side. He noticed that the compound's northern edge was also part of the city's defensive perimeter, which included fighting bunkers and lots of concertina wire. The bird then flared out and landed gently. After mouthing a "thank you" to the pilots, the two threw their bags onto the tarmac and clambered out of the Huey, keeping their heads low to avoid the still whirling blades.

They were met at the pad by Sergeant Jimmy Rocatta, the third member of the little team of advisors to the 11th Battalion of the Vietnamese Rangers.

In many forward units, officers were not saluted. The thinking was that the NVA weren't stupid and would take shots at anyone who received a salute simply because that meant the recipient was an officer. He didn't really care that Rocata didn't salute, though he remembered that Woziak had done it. DeAndre looked as the young man walked away and noticed the faded uniform and the bright new sergeant's stripes on Rocata's sleeves. Somebody who had been in Vietnam awhile and was also a brand-new NCO. DeAndre also noticed part of a tattoo on his chest near the top button. That would be a recent addition as recruits were not allowed to have tattoos.

DeAndre's new home and office was spartan. In the little "headquarters" building, there was a small office with a typewriter, a day room with TV, and two sleeping rooms. He got the nearer of the two, while the two NCOs shared the other room. The steel frame bed with the wire springs was okay; in fact, it was plush compared to the sleeping arrangements of his previous two tours. The little building was heavily sandbagged, with a heavy metal plank ceiling. Just outside the entrance was a very substantial bunker that could withstand large explosions.

The bed had a poncho liner to use as a light blanket. The three men shared using the "solar powered" black 55-gallon drum filled with water perched atop a wooden stall shower, the water warmed by the sun. Shaving could be done in the shower or the simple water basin in their sleeping rooms. DeAndre had three sets of fatigues with him and two pairs of boots. Woziak advised that there was a Vietnamese woman who also served Thiếu tá Hoang and the battalion second in command. She would wash, iron, and starch the uniforms as well as keep his boots polished. When told all this, DeAndre smiled to himself. He'd never worn polished boots before in Vietnam.

The uniforms had the Ranger patch on the left sleeve, the stylized snarling head of a black panther superimposed over a large white star. When not on operations, their headgear was a distinctive maroon beret with a badge containing a winged arrow in a wreath. In battle, they wore steel helmets, also with the black panther painted on the front. U.S. advisors assigned to Vietnamese Ranger units wore the same distinctive uniform as their counterparts.

DeAndre was tired. He'd flown across the Pacific, had endured a long ride on very noisy aircraft, and had met a lot of new people.

Wozniak looked at his boss and figured he must be on his butt. With all that travel over 36 hours, he would be tired, to say the least.

"Six, how about some chow, then you can catch some shut-eye," he said to his boss. "When we're out in the weeds, we eat what the Vietnamese eat, but I can usually scrounge some GI stuff when we're here."

And with that, he produced a choice of LRRP rations to choose from. After perusing the dehydrated meals of beef hash, chicken stew, and spaghetti with meat sauce, DeAndre pointed at the stew. Wozniak had already heated some water, so all he had to do was pour some into the bag and stir it. He probably ate too fast, but he was so tired, all he wanted to do was sleep.

DeAndre woke sometime during the night. It was hard to tell the time as there was no fancy lit clock. He heard a vaguely familiar sound but couldn't quite place it. It wasn't threatening. It was just peculiar and irregular. He was just about to slip back into sleep when recognition came. It was the sound made by a lizard that lived in the trees. They were universally known to all Americans who ever heard them as "Fuck You" lizards.

For all the world, that's what the lizards' call sounded

like; the epithet of disrespect used by most Americans at one time or another. It's not that the lizard sounded like a human, but rather that their call seemed to be the word. Before sleep befell him, he knew he was really back out in Vietnam.

CHAPTER FOUR

Waking with the dawn, DeAndre took a chilly shower because the sun hadn't had a chance to warm up the water yet, then put on a fresh set of jungle fatigues. He disliked the fact they were brand new. He was no FNG, but the unbleached look of the cammies said he was. He'd never worn camouflage uniforms before. He remembered sarcastically referring to them as "flowers" during previous tours.

As he walked outside, he saw that both of the NCOs were up and around. "Good morning, Sir, " said Sergeant Wozniak with a large smile, while Rocata stayed silent. "We usually go up to the Ranger mess hall for breakfast. The Vietnamese eat a breakfast that's a lot like ours. You know, stuff like bacon and eggs. But, no grits, sir. One additional thing they have that I recommend is the noodle soup. Try it. I think you'll like it."

"They have coffee?"

"Oh yeah—coffee that will knock your socks off. Go easy with it until your system gets used to the much higher level of caffeine."

The three of them walked a quarter of a mile or so to the mess hall. DeAndre loaded up his plate with eggs, poured himself a cup of coffee, and decided to try the soup too. They found a table, and the conversation continued.

"Meeting the major is on your agenda today, sir," said Wozniak. "I don't know exactly when, but his Sergeant Major promised to give me a heads up well before he's ready for you."

"Give me some intel on him."

"Sure. He speaks enough English so you can communicate with no major problem. Just be aware his pronunciation isn't always the best. For instance, do you know what a YerOp'ian is?", as he stressed the second syllable.

"What? A YerOp'ian?"

"Yeah," snickered Wozniak. "You know—people who live in Europe. The Vietnamese often put the emphasis on the wrong syllable. Most of them learned English from other Vietnamese, so their pronunciation is not all that good. However, Thiếu tá Hoang's second in command is Đại úy Pham. He spent six months at FortBenning, so his English is pretty good, including his pronunciation."

"Okay—I give up. What's a Đại úy?" came the slowly spoken pronunciation of die wee.

"Same same you, Sir. A Captain."

"One more thing, Sir," continued Wozniak, "Don't be surprised if Hoang looks a bit startled when he sees you. I don't know if he has ever met a Black man before, but I know he hasn't seen any since I came here. They seldom see any Americans except their advisors."

Breakfast finished, the three ambled back to their hooch, giving DeAndre a chance to get his bearings a bit. Vietnamese Rangers in groups of twos and threes walked by, M-16s slung on their shoulders and walking confidently. He noticed good morale and that they all took their military courtesy seriously and saluted, though he could not understand what they said.

He returned their salutes crisply, with a strong, "Good morning, gentlemen" with the full knowledge that they had no idea what he said either. In many ways, the Ranger camp was not that different from an American one. There were a few wooden frame buildings that looked like they were used for administration, and there were a lot of GP Medium tents, each with a wooden floor and each with a nearby shower. Every place that was inhabited was well protected with sandbags. It was a bit more spread out than an American base, and it also did not have any artillery pieces. The Rangers had no organic artillery, and had to depend on regular ARVN units for support. The camp had a helipad big enough to accommodate a Chinook or two Hueys, but only Vietnamese aircraft used it. Americans used their own, elsewhere in the city. The whole camp was surrounded by concrete fighting bunkers. That was a huge difference from the Americans who never used concrete because they knew they were going home someday.

But the Vietnamese were already home, so they built things more permanently. DeAndre also noticed there were trees inside the perimeter which meant the Vietnamese didn't bulldoze the entire interior of the camp. He also noticed a two-story building and figured it was probably the battalion headquarters.

He was surprised at the presence of regular ARVN

troops on the compound. Wozniak informed him that there was a logistics unit and a corresponding security company assigned. They secured the perimeter when the Rangers were out on operations.

When a runner advised Wozniak that the major was ready to see Captain Washington, he was led to the second floor of the building where he was escorted into Thiếu tá Hoang's office. Stepping in front of the big oak desk, DeAndre came to attention and made a sharp salute, while saying, "Good morning, Sir."

Major Hoang casually returned the salute as he came around his desk, then motioned DeAndre towards a comfortable chair. Sitting across from him, the major crossed his legs and seemed to relax, which put the American at ease too. If he was startled by DeAndre's race, he didn't show it.

"Welcome to little camp in Kon Tum," said Hoang with a very heavy accent. "You have no insignia on uniform except name and 'US Army.' I give you patch for 11 Battalion of 2nd Ranger Group."

Reaching over to Washington, he handed him several blacked-out insignia of the Ranger patch with the numbers 11 and 2 imposed. "I have servant sew these on for you. I know you airborne and Ranger in American army. I have those sewn."

Washington was impressed that Major Hoang had done his homework. "Thank you very much, sir. I'll wear the patch with honor."

"My English not so good."

"It is much better than my Vietnamese," rejoined De-

Andre. That seemed to put the major at ease. "But please forgive me if I sometimes have to ask you to repeat yourself."

The conversation then moved on to some brief sharing of past assignments before they began to talk about operational issues. At the top of Thiếu tá Hoang's list was wondering how much he knew about bringing in American jets for close air support.

"No problem, sir, I did that numerous times as a Company Commander with the 1st Cav. Once I go through my own command channels, I will contact the assigned Forward Air Controller. He flies around in a small single engined propeller plane, but he also translates what I say into instructions for the jet pilots. He will have one or more aircraft assigned."

"Ah, good, Captain. Who decide if we ask for snake or nape?"

DeAndre realized he'd been had a little. Major Hoang knew all about using the FAC and had watched previous American advisors do it. He just wanted to be sure he didn't have a newbie. His question about "snake or nape" had to do with whether the major decided to use 500-pound bombs that penetrated bunkers or use napalm. Hoang was trying to find out if he had an arrogant American with him who really wanted to run the whole show, or if he was truly an advisor.

"Oh, you do, Sir. That is a command decision."

Just the answer Hoang wanted to hear.

"Also, Captain. I not want VNAF planes," saying the letters like "Vee Naph." Hoang meant he did not want

Vietnamese Air Force planes to provide close air support. Like other ARVN commanders, he didn't always trust his compatriot's flying skills and precision.

"I'll do my best, Sir. But you know there are fewer and fewer American planes in Vietnam. As American Army units leave, so do Air Force units. If I cannot get American planes, I will let you know and you can decide what you want to do." DeAndre just wanted to reinforce that it was January 1972, and the number of Americans was declining swiftly.

"Friday, we take two companies to villages south, just to, ummm—how you say, 'Look around.'"

"I think you mean recon, Sir."

"Yes, we will recon. There are some villages there we check. Good way to find out what enemy do. How you say—we do this often."

"Just a routine operation, sir."

"Yes—routine. Maybe we have big operation soon."

Walking back to his hootch, DeAndre liked what he saw but reserved final judgment until he'd seen Hoang under fire.

Thursday evening, SFC Wozniak chatted with the captain about how they would operate in the field. Rocatta would carry the captain's radio. "Don't worry, Sir. Rocatta's a damned good RTO. He was a company radio man with the 101st before his battalion went home."

Wozniak went on to say that in most units, the Vietnamese commander would assign a radio operator, but Washington's predecessor had talked Hoang out of doing that, fearing the man might report back to Hoang about what the Americans were saying. "I don't know what Captain Merrick said to him. No Vietnamese would normally allow an American to carry a radio because it would mean he was a bad host and the Vietnamese commander would lose face."

But he did find out that when in the field, the Vietnamese would assign a "batman" to the Captain to cook his food, dig a foxhole, act as bodyguard, or anything else required. An officer's batman was an old European tradition of having an enlisted man take care of all the trivial matters for officers to ensure nothing distracted them from leadership duties.

Wozniak was ostensibly supposed to stay with the battalion Sergeant Major while on operations, but in reality, he went wherever he wanted to go. He packed his own radio and could talk with the captain whenever a need arose.

The battalion planned to be in the field for only a few days during the recon and search mission. It would be very hard to carry enough of the canned c-ration meals, so they decided to carry LRRP rations. "There's a fair number of spring-fed streams in this area and the map shows a few in our AO. That should keep us in water for both drinking and fixing LRRPs," said Wozniak as he handed out the plastic wrapped packs.

= = = = = =

Breakfast was before dawn Friday morning. By seven, the two companies were formed up in full battle gear and awaited the trucks to transport them the 30 miles or so to the first village. DeAndre knew the Vietnamese Army didn't have enough Huey helicopter pilots to provide lifts for all operations, so they often had to revert to vehicles on the road.

Narrow and winding, the dirt road was just wide enough for two trucks to pass. Being on roads was a bit spooky for DeAndre who was used to flying into combat where the enemy couldn't shoot at you.

Worse, he was riding in the back seat of a jeep with Major Hoang. Two tall whip antennas made for a flashing neon sign screaming "Senior Officer Here! Senior Officer Here!"

"Not worry, Đại úy. Units check road for mines. We safe."

Even as he was saying, "Yes, Sir ", Washington was looking nervously at the thick forest on both sides of the road. He was especially observant as the convoy approached a bend in the road—a place where the enemy could set up an ambush and cover the entire convoy.

DeAndre wondered if his radio operator had any thoughts about ambushes. "Not so worried." The captain told himself that Rocatta was a man of few words.

True to plans, no bullets came their way. As the convoy approached the first village, he saw a few trucks stop and some of the troops disembark, then double-time into positions surrounding the houses—if they could be called houses. More like huts made of small tree limbs or bamboo with a roof of dried grass, each one was elevated off the ground on thick posts. Some few had big pieces of corrugated tin covering the walls. These were not at all like the Vietnamese houses DeAndre was used to seeing. There was also one huge house, very long and built up on stilts. The place screamed poverty.

Nor did the people look Vietnamese. Of darker skin, they lacked the eye structure of most Asians. DeAndre was startled to see that some of the women were bare-breasted, but most wore black garments decorated with colorful zig zag patterns and their little children wore only a shirt and no pants. Washington assumed they didn't use diapers.

Many of the men wore only loincloths, while others wore western style shirts and pants. All the villagers wore dour expressions. Some of the women and children showed fear while some men straightened their backs in a show of defiance to the Vietnamese soldiers.

"Jarai?" asked DeAndre of his radioman.

"Yeah."

Assuming Rocatta's short answer meant he didn't know much about the mountain people, he tucked it in his mind to ask Wozniak later that night.

With the village sealed off, the rest of the troops exited the trucks and began to systematically enter the houses, looking for weapons and any other items hidden from the Vietnamese. Following at a discreet distance, DeAndre observed Thiếu tá Hoang as he supervised the search. Fifteen minutes after the Rangers arrived, a Vietnamese couple approached the major. Nicely dressed and smiling, they settled into a seemingly pleasant conversation with the commander.

Seeing that, Rocatta turned and walked away without saying anything. Catching up to him, DeAndre asked his radioman what he was doing. "Ask Sergeant Wozniak later." The answer annoyed the captain, but he decided to drop it for now.

DeAndre wandered a bit further away from the major and just watched the troops at work. They checked out rain barrels looking for underground shelters as well as checking food storage areas. The occupants squatted passively outside, sometimes holding back their curly tailed yellow dogs which barked and snarled at the soldiers. If these troops were dishonest, there must not have been anything worth stealing in that house as he didn't see them coming out with anything.

There was shouting behind them. DeAndre turned to see two Rangers roughly shoving a young man from another house, poking him in the ribs with the butts of their M-16s. The kid and the soldiers were all screaming, but to no avail as neither understood the other's language. Once in the street, the soldiers butt-stroked their prisoner in the groin who then collapsed with a groan.

Đại úy Pham had been watching both the Rangers and DeAndre. "By law, all men of military age must carry a card with their military status," said Pham. "The troops found him in an underground shelter, and he has no ID card."

A bit startled at the Vietnamese captain's excellent English, DeAndre said he understood and mumbled something about American draft dodgers, which made the US trained Pham smile.

The rangers handled the kid a bit too roughly by DeAndre's standards but let it go as he was not yet familiar with the culture. He thought of his reputation of wanting respect, not just fear, from the troops, but hid that from the Vietnamese.

But, as the search progressed, he noticed that same level of near-brutality towards women as well as men. In one case, he saw two soldiers shove an older woman to the ground and kick her when she strongly objected to them taking something from her house. He couldn't tell what they took, and she appeared unhurt, but thought it was a bit too much to shove and kick the gray-haired lady.

He'd noticed the battalion commander was still talking with the Vietnamese couple, and it seemed like a very pleasant conversation. They all smiled and laughed lightly, seemingly unaware of the rough treatment the soldiers were giving the Jarai.

The search ended after two hours or so. Thiếu tá Hoang cheerfully said goodbye to the Kinh couple before getting into his jeep while his troops clambered into the trucks. DeAndre noticed a few of the soldiers had additional items that they didn't have before coming to the village.

The sun was approaching its apex when the column halted near an open area. Pulling into the field, the vehicles circled up, making a nice and neat defensive position with good fields of fire. Setting up cooking fires, the troops soon had lunch ready. An orderly brought food to Thiếu tá Hoang, as well as to the Americans. It was a simple noodle soup with a nice sandwich of a baguette and some kind of mystery meat. Both were delicious.

But then, to Captain Washington's great surprise, many of the troops hung hammocks while others stretched out on the ground or truck bed. Even the officers napped. Only a few men stayed awake and on guard.

"Welcome to Vietnam, Sir. Almost all of them take a siesta after lunch. I wonder if the enemy does it too because I've never heard of a South Vietnamese unit being attacked during the noon nap time." Wozniak explained that if they were in a city or town, most likely the troops would also have had a beer—or two—or three—before they slept.

Remembering that the infamous Tet Offensive of 1968 had started when the southern troops thought the enemy would not attack then, DeAndre was too nervous to sleep himself. He cradled his M16 and kept his eyes peeled.

The afternoon was spent the same as the morning. The soldiers, some of them inexperienced, some older and more jaded, went through the same procedure that they had just done in the morning. They searched a village as the major talked with the few ethnic Vietnamese in the village. This time, there were two couples and he talked to all four of them at once. Washington stood back, deeply curious, but not daring to watch. He did notice one of the men hand the major a red envelope. He knew that even if he could have eavesdropped, he wouldn't have had a clue what they were saying. The search ended the same as the first with the soldiers carrying off a few small things and the major leaving after exuberantly shaking hands with the two couples, a broad grin on his face as he got in the jeep.

The battalion found another open field where they set up an overnight defensive perimeter with foxholes, machine guns, and claymore mines in place. They ate dinner before the sun set, then the men manned their positions just before last light, three men to a hole, sleeping on low-slung hammocks mounted between stakes or on the ground. He noticed there were no air mattresses, a policy he wholeheartedly agreed with as the rubber in them made a lot of noise.

It was a good time to ask SGT Wozniak about the mountain people. "There might be some Bahnar around here, but don't ask me how to tell the difference between them and Jarai. I can tell you there is no love lost between the tribes and the Vietnamese." DeAndre mentally filed this information next to LTC Arredondo's comments about the enmity between the ethnic Vietnamese and the hill tribes.

"Rocatta mentioned other people. He used the word 'Kinh.'

"Ethnic Vietnamese. By the way, none of the tribes we call Montagnards speak Vietnamese. They all have their own languages. Oh sure, you will run into the occasional tribesman who speaks Vietnamese, but the average peasant doesn't."

DeAndre also asked about Thiếu tá Hoang meeting Kinh in the village, along with the fact that it seemed like Rocatta was steering him away from watching the Vietnamese major. "I'm not sure why Hoang doesn't want us to know he is talking to village Kinh, but Rocatta led you away because if you hadn't turned your attention away, somebody would have come over to distract you to be sure the major had privacy."

"Wonder why he doesn't just go into their house if he doesn't want me watching."

"Beats me, but my guess is the Vietnamese don't want the soldiers to know they have valuable stuff in the house. The Rangers have a bad reputation for just taking what they want from houses, though they usually leave the Kinh houses alone."

Sometime during the night, as he lay awake, he saw the faintest little dot of light out on the perimeter. In less than a minute, he heard a hushed yowl as the soldier who had violated light discipline by putting match to cigarette was kicked in the face by his sergeant.

"Hmmm" thought DeAndre. "Different army than mine."

The next two days were the same. Truck to another Jarai village, search the houses with the major talking to Kinh, and a few small items taken from the homes. Sometimes the brutality was fierce, as in the interrogation of one young man who was dragged out of his house and beaten badly by the Rangers before they took him into custody after he lost consciousness.

Back at base camp, the trio settled into their hooch and broke out the Biere Larue. After commenting on how good the cold beer was, the senior sergeant and the captain lapsed into talking shop while Rocatta wandered off, seemingly to listen to Armed Forces Radio.

"Is it me or does that kid not like me?" asked DeAndre.

Wozniak gave his boss a startled look, then paused a split second before answering. "Oh, I wouldn't worry about it. He's a good soldier, but he's also still a teenager. Just moody."

DeAndre said nothing, but something inside told him there was more to it than moodiness.

Changing the subject, Washington said he had noticed that the only staff car on the base, along with a couple of trucks, had left fairly early in the morning and did not return for a couple of days or so. He asked Woziak what that was about, but he said he didn't know. Eventually the senior NCO opined that his suspicion was that the Thiếu tá was talking to the Kinh in the villages and taking money from them in exchange for troops moving them safely away from the Jarai villages and into Kon Tum and/or Pleiku.

"There's no way I can prove that, Captain—not without overtly following the staff car and trucks. Somehow, I think that would piss off Major Hoang."

"Yeah, I understand it's just your opinion, but there's one hell of a lot of likelihood of that theory being correct. Colonel Arredondo seems to think Hoang is crooked in some way—maybe that's it. That just makes me worry about him more."

"What do you mean, Sir?"

"I keep wondering why we don't get ambushed while

we're trucking to the villages. You don't suppose Hoang buys off the NVA, do you?"

Wozniak suppressed a smile at that idea. "I doubt that, sir. When you see Hoang in a fight, you'll see what I mean. He's no traitor, nor will he do anything to hurt South Vietnam."

"Then why don't we get ambushed on the road?"

"Well, sir—the NVA are in strange territory just like we are. They don't live around here and they would have no reason to trust the locals to help them. During my first tour, I was with the 199th Light Infantry Brigade. We were fighting the VC then, not NVA. The local farmers knew the terrain very well indeed. They knew where to set up booby traps because they walked the area every day while out farming. The NVA can't do what local VC can do. They're strangers too."

"No VC around here?"

"Not that I've ever heard of, but I'll ask the staff intelligence shop the next time we're in Pleiku. But even if there are, few would be locals. The Tet Offensive back in 1968 did a job on the VC. A lot of them were killed. Today, many units are VC in name only with their troops being mostly northerners."

Sergeant Rocatta had been listening from afar. With a gruff tone of voice, he chimed in. "There's no damned VC around here, unless there's some in the local Ruff Puffs." With that, he turned up the music on his radio, as if to say he would let the two old guys talk their own silliness.

Both DeAndre and Wozniak glanced in Rocatta's direction. They both knew he was talking about village Regional Forces and Popular Forces that were kind of a village militia. Wozniak then replied, "Not around here. You won't see any Ruff Puffs in the Montagnard areas. I think the closest unit is just outside Pleiku. The Green Beanies run something called CIDG. Stands for Civilian Irregular Defense Groups and they are mostly hill tribe members initially trained to protect their villages. It didn't take long for them to get a more traditional mission. Under American officers, the CIDG became damned effective fighters. There's no problem with them being VC. More likely, the troops don't like the Kinh people or government. Back in 1964, a bunch of Rhade CIDG troops rebelled in Ban Me Thuot. Killed a bunch of Kinh civilians too, until a Vietnamese division put the mutiny down. But today, that's considered the birth of FULRO."

"Okay—tell me more about FULRO. The colonel didn't tell me much," said Washington.

"It's a movement among the tribes to create an independent country just for the hill tribes up here in the Highlands. The Vietnamese government hates them—big time hates FULRO."

After learning that an experiment was tried and failed in converting CIDG teams to Ranger battalions with Vietnamese officers, the talk drifted back to discussing the search missions in general and Thiếu tá Hoang in particular.

Sergeant Wozniak said, "Ya know, I asked Captain BeShear, the advisor to the 2nd Ranger Group operations officer, if Hoang is instructed to do those village sweeps or if he just makes them up. Seems that those operations are totally freelance. Hoang just files a report after the fact."

Rocatta came back for another beer, listened as the conversation shifted again, then went back under the tree. There had been some talk that there might be another major offensive like the infamous 1968 Tet offensive. But by February 7, the last day of the Lunar New Year, it had become clear that an offensive was not going to happen then.

All the same, intelligence reports told of an influx of enemy troops in southern Laos and northern Cambodia on the border with Vietnam had become more alarming. The additional troops seemed to be preparing for something big, an offensive that could come when no one expected it.

The two talked a while longer, discussing the intelligence reports. One bit of information was that the enemy had modern Soviet tanks. The Rangers were light infantry with no heavy weapons nor even their own artillery. If they were ordered into a fight against tanks, they'd be slaughtered. In the far distance, they all heard the rumble of artillery. They didn't know if they were American or ARVN guns, but it didn't really matter.

The cache of empty beer bottles had grown to a good-sized pile. The conversation quieted as the two thought of getting some sleep. "Well, I'm ready to hit the rack," mumbled the captain as he ambled towards his room. "G'night, sir," said Wozniak, loud enough for Rocatta to hear, who then chimed in with a quick "G'night." De-

Andre mumbled a barely coherent, "G'night, gentlemen," then headed for his bunk.

Bookshelves in his room from previous occupants held copies of *The Hobbit* and *The Lord of the Rings*, which reminded him of his first tour as a lieutenant when that was the only reading material he had for months. As he drifted off to sleep, he tried hard not to imagine himself as Aragorn, the Wanderer who became king, even though he doubted his own ability. Perhaps that is part of why Aragorn was heroic—just an ordinary guy who humbly did heroic things. The conversation inside his head told DeAndre he was no hero. He really wasn't heroic. He also knew at some level that women were attracted to him. He liked his reputation.

He fell asleep wondering if the woman of his dreams would think him "heroic." Susan knew he was no hero, but she knew he was damned good in bed. DeAndre enjoyed an erotic dream of her before going into deep sleep.

Morning came with the knowledge that there would just be another boring operation. Just trips to the villages for "recon", thought DeAndre. The only variance was that Thiếu tá Hoang announced the entire battalion would participate in the cordon operations to roust out the Jarai villages.

Colonel Arredondo had told DeAndre that when an entire Vietnamese Ranger battalion moved in the field, the Commanding Officer would travel with two of the companies while the Executive Officer moved with the other two. The 11th Battalion hadn't moved as one unit since he'd been there, but during the next village search the Rangers did just that.

"Do you have a good relationship with Đại úy Pham?" asked Washington of Sergeant Wozniak.

"Yes sir. He seems to be a very good officer and is actually quite personable."

"When we move out with the battalion, I will stay around the major as I usually do, but I'd like you to stay with Pham rather than wander around as you usually do."

"Okay. Any reason why?"

"If I don't travel with the major, he will take it as losing face. He might also be suspicious of me—worried I might be snooping around. But, it would be expected if you were to work with the Đại úy. Don't overplay your hand—just kinda buddy-buddy with him. I am curious if he would ever tell us what Major Hoang does when he talks with the Kinh in the village."

The next morning was same-same, but different. Same hassling of the Jarai, different village, but with twice the number of troops as usual. There was no road to the village, so the troops had to hike for a half hour to get there.

As instructed, Wozniak went with the two companies headed up by Đại úy Pham. This place was much larger than previous villages and seemingly was a bit more prosperous. There were three long houses, again very long and built up on stilts. Many families would live together in them, except for a smaller one which was reserved for young single men.

The village included a tall pyramidal building atop a platform. There were also a few concrete block homes huddled together in one area not far from some sort of commercial building. As DeAndre looked at them, he saw people emerge and walk towards Thiếu tá Hoang. They were Kinh. His back to Washington, Hoang warmly greeted the group of about twenty people. Washington watched until he saw the major turn to begin walking, then Washington turned aside.

He also noticed the tang of smoke in the air. Noticing that nobody seemed particularly alarmed by the possibility of fire, DeAndre wandered off looking for some kind of explanation.

"You sure you want to be so far away from the Rangers?" asked Rocatta nervously.

"Yeah, I guess we ought to stay near them, eh?" grinned Washington. "I was just trying to find out where the smell of smoke came from."

"You don't have to be up here very long to know the answer to that," said the young sergeant. "All the 'Yards are farmers and they do slash and burn agriculture. What you smell is an area that was recently burned. Ask Sergeant Wozniak to explain more. He's lived up here a lot longer than me."

"Sounds good. Let's get back to the main body."

Thiếu tá Hoang must have noticed the absence of Washington and his radioman and gave them a hard and inquisitive look when they returned. He was still talking to the group of Kinh but made sure the Americans realized he knew they had disappeared awhile. At the end of the day, as everyone mounted their vehicles, Hoang turned to Washington and asked, "Where you go?" with a slightly belligerent tone.

"I was wondering where the smoke came from, but Trung si Rocatta told me."

The answer seemed to satisfy the major and he turned to talk to his driver and used the radio to talk with his second in command about the other two companies.

While eating LRRP rations that evening, Washington asked Wozniak about slash and burn.

"It's how all the tribes grow their food. They all farm. What they do is cut down all the trees and brush in an area, then let it sit for a year or so. By the time the cut trees start to rot, they light all that on fire. The ashes nourish the soil very well and the people can grow their food on it for five years or so until the land starts to play out.

"They just complete the process over and over. The farmed area will grow back in about six or seven years or so, and the village just repeats the process all over again. Sometimes slash and burn is called swidden.

"The Vietnamese hate them doing it and try to get them to stop. But the Jarai and all the other tribes know what the Kinh really want. This mountain area is great for growing both coffee and tea. The Vietnamese figure they can get rich off growing those crops."

"Any association between the coffee and tea growing and the Kinh you see in almost all the villages?"

"Oh, yeah. The government's official policy is to assimilate the tribes and take over the land. They make no bones about it. Many Kinh are either Catholic refugees from the north or have fled the fighting in their home villages. Most of them build the shacks you see in Kontum and Pleiku. The government pays the big city people to settle in this area, particularly in and around the Montagnard villages and give the Kinh a toehold.

"Have you noticed that the Kinh don't seem to have any way of making a living other than a few stores? They obviously are not farmers. Gotta make money some way until they get their hands on the land and start planting coffee or tea."

"There must be some connection between those people and Thiếu tá Hoang meeting with them when they search the villages," pondered DeAndre. "I also wonder about the red envelopes. Why red?"

"There's a connection, alright. I just don't know exactly what goes on between him and the people he talks with," finished Wozniak. "But, from my experience during my first tour when we were working amongst the villagers, I can tell you the red envelopes are something usually seen during the lunar new year. Parents use them to give money to their kids. Red envelopes are also used for, shall I say, non-business dealings."

"You mean bribes?"

"Yes, but it's not always bribes or anything unethical."

And with that, the dark of the tropical mountains set in with the usual fantastic display of the nighttime stars. The three climbed into their hammocks, all slung near their foxholes, and soon fell asleep. But before doing so, DeAndre kept wondering when they would see real action. He'd been there for quite a while and all that happened was that the Rangers were screwing with the people in the Jarai villages.

And DeAndre was tired of seeing women manhandled, young men beaten to a pulp, and things taken from homes. The discrimination and intimidation were fierce, reminding him of stories his grandparents told him about living under Jim Crow.

CHAPTER FIVE

On a bright morning in late March, Washington got the change he wanted. The sun was just cracking through when a runner came from Thiếu tá Hoang reporting that the rest of the operation was canceled and the battalion would return to base camp as soon as possible. The captain went to his radio to ask his boss if anything was going on, but Pleiku was out of range of his field radio. The batman came to gather up the Americans' gear and DeAndre walked the short distance to Hoang's command post.

Anticipating Washington's query, the Ranger commander said, "Knowledge many enemy in Kampuchea and Laos. Maybe big fight come." Now DeAndre really wanted to talk with his boss.

A day after their return to base camp, Hoang sent a runner to the Americans asking that Washington join him for a staff meeting at 10 A.M. that morning. DeAndre entered the battalion headquarters, finding his way to the operations room adjacent to the major's office.

"Good morning, sir," said Washington as he approached his mentee.

"Good morning, Đại úy." With that, Washington was instructed to sit next to the major during the briefing. "I give you interpreter so you know," a moment before

Washington was about to ask how he would be able to understand. With that, a pleasant looking young man in civilian clothes smilingly approached DeAndre and introduced himself. "Good morning, Captain. My name is Nguyen The Anh, but please just call me Anh." His baby face smiled as he extended his hand. "I studied in America for four years, in case you are wondering."

"Ah, that explains the very good English," DeAndre smiled back. "Pleasure to meet you, Anh."

"But I'm not a soldier, so I hope I can tell you the military gibberish correctly. By the way, I am the son of the mayor of Doan Ket, a village outside Kon Tum town. He sent me to college in America and now I'm back."

DeAndre was amazed at Anh's impeccable English. He had only the very faintest accent. "I don't think the fact that you're not a soldier will make any difference at all." The two drifted into a comfortable conversation as they got to know each other.

Soon, the Vietnamese operations officer got everyone's attention. After a few moments, Anh swiftly and easily translated for DeAndre. "Intelligence believes there is a large—and growing—enemy force across the border in Laos, right near the intersection with Kampuchea, or Cambodia, as you call it. Even new Soviet tanks had been spotted."

The battalion was to be a part of a large Ranger and Airborne operation west of Kon Tum. Intelligence indicated the enemy was readying for a big push, coming in from the Ho Chi Minh trail. The battalion's assignment was to attack as soon as the Communists crossed the border while regular ARVN units were ordered to act as a blocking force to prevent the enemy from escaping.

With a nod to Anh, DeAndre thought to himself, "So now I finally find out what the guy will do in combat. We'll see how he handles it when things go wrong—as they inevitably do in a fight. I'm going to find out the real Thiếu tá."

He was pleasantly surprised that the Rangers used the same Five Paragraph Field Order he had been trained to use. The Operations Officer apprised everyone of the situation, then a young First Lieutenant provided the intelligence background, including the weather forecast. The Operations officer took over again, giving a short crisp mission assignment. He continued by detailing instructions on how the battalion would accomplish that mission, right down to the mission for each of the companies.

The staff officer in charge of logistics let everyone know how food, additional ammunition, and other resupply issues would be handled. Last was information about radio frequencies, call signs, and other communications information.

There was one huge surprise for Captain Washington. While the plan called for trucking the troops to a jump-off area, the full details of where the battalion was going and when they would be picked up by Chinook helicopter were intentionally not covered. The Vietnamese were worried about spies in their midst, so such details were omitted until the last possible moment.

DeAndre was dumbstruck. How do you plan for artillery support if you can't tell the guns where you're going?

How do you plan your logistics without knowing how long you would be in the field? Since there weren't many helicopters, how did you plan for tactical movement to catch the enemy? How would they coordinate with the regular ARVN unit they were working with? How would they get maps to the eventual AO?

But—that's the way it was. He remembered that Colonel Arredondo had warned him that he was not in command—he was an advisor. He later discovered that this was not an unusual way for the Rangers to handle big operations that involved multiple units.

The plan called for the battalion to move the next day. Trucks would take the troops and a large number of supplies to the coordination area where they would receive the rest of the orders. Washington realized the Rangers too were cognizant of the fact that getting those final orders so late made logistical planning hard, so they compensated as best they could by taking a lot of ammunition, food, and other supplies to the jump-off where they would be guarded by a platoon of regular ARVN troops.

DeAndre also noticed that the movement out of the administrative area was by Chinook helicopter, which meant that the first movement would be into a secure area. The giant twin rotored birds were way too big a target to be used for combat, so it must mean the Rangers would ride 'Hooks" into a staging area, then the actual helicopter assault would be by much smaller Hueys, which were better suited to combat operations.

Anh tried hard to keep DeAndre informed, but there was a lot of excited chatter among the Vietnamese planning the operation, but he got the gist of it. After a while, Hoang turned to Anh, said something to the translator, then went back to work.

"I won't be able to go with you on the operation," he said, "but Thiếu tá Hoang told me he will assign a translator to you. I am a civilian and too much of a risk."

While that news disappointed Washington, he was not surprised. He thanked Anh profusely for his help and the two parted when the briefing ended.

Back in the little American headquarters, DeAndre briefed the other two. "No surprise on this, Captain. Most of the time they leave out final destination location and operational times. Some units have had problems with the enemy throwing them a hot welcome party when they arrive," said Wozniak. "However, this does sound like we'll be out in the weeds awhile and we won't have vehicles, so we'll have to hump everything. We won't be able to take enough LRRPs with us, so we'll have to eat Vietnamese food during the trip. Don't worry—they'll have enough. I'll let the Sergeant Major know."

Settling into his bunk that night, DeAndre knew he would be sleeping in a hammock or on the ground for a while. Smiling, he snuggled down into his poncho liner. As sleep overcame him, the memory of his mother on the day he'd left home after the church service came to mind—and he remembered her fear of him going into combat again. But he was more concerned for his mother than his own safety.

Finally—a real operation. Washington felt, but did not welcome, the first small surge of adrenaline. Thiếu tá

Hoang seemed to be pumped up too. As promised, he assigned an interpreter to Washington who couldn't fit into the back seat of the jeep with the two Americans but was in the vehicle that followed.

Heading north on Route 14 towards Dak To, the convoy weaved its way through forest land, some of which had been turned into farmland, and some very small hamlets. The road slowly curved towards the west. Just after they passed the Dak Poko river they continued west on Route 40. They were heading to the small town of Plei Re near the border with Cambodia.

But to get there, the twisting dirt road traversed some very high mountains with jungle growth right at the edge of the narrow road. Everybody felt tense, sensing the possibility of enemy ambush. Every truck had a machine gun mounted behind the cab, and every gun was manned.

The bumpy, dusty ride to the admin area took a few hours, but when they finally exited the mountains, Hoang turned and told DeAndre they were near the destination. He was happy he would soon be out of the bouncy jeep. He could see the town, but before the convoy reached it, Thiếu tá Hoang indicated they would set up in the open field at the intersection of three roadways.

DeAndre turned to look at the trucks and was delighted to see how efficiently they moved into a circle. The troops dismounted and assumed a very good perimeter defense within minutes. Both the commander and his advisor heard a low distant rumble coming from the west in Cambodia. It was an Arc Light—a B 52 strike that dropped over 90 tons of high explosives, mostly 1,000-pound bombs.

Washington had been near one of those strikes in 1969 when his company couldn't be extracted in time and his men were inside the danger zone. It had been a scary experience as the company was ordered into an open field and everyone was instructed to lie on their stomach. The ground literally moved up and down—and did so with such ferocity that everyone had bruises on their chest for days afterwards. Huge chunks of shrapnel whizzed through the air, and one of them neatly lopped off the top of a tree on the edge of the field. DeAndre knew full well that no one in the kill zone across the border would be alive.

As the battalion settled in to await orders that would send them into action and to coordinate with the attached ARVN battalion, Hoang's radio crackled. He was ordered to stand down and return to base camp as quickly as possible. There would be no heliborne operation. Hoang told Washington the news himself, then turned to issue orders to his own subordinates. DeAndre was mildly disappointed, yet he also had forebodings that a really huge fight was coming.

He was right.

Back at base camp, he radioed Col. Arredondo and got the news. A huge invasion of South Vietnam was going on, and not just in their area. Northern troops had overrun a number of firebases along the demilitarized zone that divided north and south Vietnam. The enemy had converged on Quang Tri City, just south of the border, and laid siege to it. Another incursion had started near An Loc, well south of the Highlands.

Final plans hadn't been set yet, but Arredondo said that regular ARVN units would move towards the border as well as to protect the camps at Polei Kleng and Ben Het, both of which stood between the Communist forces and the provincial capital of Kon Tum. Other major forces from outside the area would be sent to Tan Kanh and Dak To.

The Rangers would be pulled back to Kon Tum to be the area reaction force and final defense of Kon Tum. For the Rangers and their American advisors, days went by, and as they listened to reports of other big fights in other parts of South Vietnam, the accumulated adrenaline began to burn.

To help contend with the lack of action, Washington tracked down Nguyen The Anh, the translator, and invited him to the hooch. The young man was refreshing, being highly Americanized, yet thoroughly Vietnamese. He'd learned to drink beer at UCLA and took the proffered can eagerly.

The two swapped stories about being minority students at big universities, but DeAndre "won" that contest when Anh admitted there were many Asian students on campus. Slowly, the conversation changed to Vietnam. Anh had majored in Political Science and had a practical mastery of Vietnam's political system. His father being a mayor added to his knowledge. He was pessimistic about the south's future, feeling that the corruption was undermining efforts to be truly free and democratic.

One of the biggest changes had been his feelings about the hill tribes. He no longer called them "moi" and was quite vocal in his demand for better treatment, which put him at odds with his father.

"If I understand it, the people around here are Jarai. Is that right?" asked DeAndre.

"Yeah, they are. The Jarai is the largest group and easily the most numerous around here. There are settlements of them in Cambodia, too."

DeAndre was surprised to hear that a substantial number of them were Christian. "Evangelical Christians, to be exact," said Anh. "The Catholics came to Vietnam way back in the 16th century. Today, about 10% of the total population is Catholic. But the Protestants didn't get to Vietnam until 1911.

"Finding they were way behind the Catholics, the Protestants decided to go to the mountains. Bingo—that's why the Christians found among the minority tribes are usually Protestant, and that means evangelical Christians."

DeAndre learned there was a small Jarai section in Kon Tum, though it was more like a ghetto. Most were people who had been displaced by the war, their villages destroyed, or the Communists hunting them down because they were Christian. Knowing this whetted DeAndre's appetite to know more.

"Could you take me there sometime, Anh?"

"Sure—be happy to take you, but you want to be sure Thiếu tá Hoang doesn't find out. He totally hates the Jarai. You would be outta here in a flash if he found out."

CHAPTER SIX

When Anh returned the following morning, they plotted. It was agreed that Anh would go first and by himself so that the guards didn't make an association between the two. For him, he only needed to tell the guards he was going home. Besides, they were used to seeing him on his many trips to translate. A bit later, DeAndre would go through the same gate, telling the Vietnamese guard he was going to get some American beer from a helicopter crew at the airfield.

He knew that was unusual because most officers would be driven over, but Anh had told him the Vietnamese considered the American officers to be weird and would not question it.

The next day, Anh returned to the hootch. As he and DeAndre lazily walked up to the Ranger mess hall, the conversation became friendlier as Anh described how he was changed by going to college in America.

With a conspiratorial nod to DeAndre, he recognized the prejudice against Blacks in the US and saw the same in his own country with the discrimination against the hill tribes. Though a privileged Kinh living in the highlands, he no longer agreed with such bigotry.

Anh took to going to the Jarai ghetto on occasion

and even learned a little of the language. Finally, his father found out, leading to multiple screaming sessions between the two. Being an older traditional man, Anh's father threatened to send him to Saigon, but Anh knew the old man would grudgingly let him stay home because Anh was the oldest child and the only male.

After breakfast, the two walked towards the gate, but DeAndre hung back, out of sight of the guards, until he saw that Anh had successfully left the camp. As per their plan, Anh waited around the corner until the American arrived, who was feeling very exposed being a foreigner by himself.

It was a surprisingly short walk to the Jarai section. The two spent an hour or so wandering around. DeAndre was struck by how very different the Jarai lived than the Kinh. There were few concrete block homes, with most being wattle huts with thatched roofs. The smell told him that sanitation was poor. There were a few little mom and pop shops that sold the basics, but no place that sold fancier items. In other words, a place where refugees had settled. When DeAndre spotted a wooden building that displayed a cross outside, he was assured that it was indeed a church. DeAndre also noticed it sported a fresh coat of whitewash which didn't quite hide the worn exterior of the church.

Anh suggested that an hour was enough. He was a little afraid that their absence would get back to Thiếu tá Hoang, so he returned to his home and DeAndre walked back to the base camp, returning the guards' salutes.

"I wonder if they noticed I didn't have any beer with me," he mumbled to himself.

The war still seemed a long way off. The next day, DeAndre and Anh went into the Jarai area of Kon Tum again. Feeling very conspicuous in uniform, DeAndre had realized he needed to wear civilian clothes. He wouldn't be able to hide he was an American, owing to his dark skin, but it would be better that people didn't know he was with the American army.

This time, Anh smuggled out some of Washington's civilian clothing, and DeAndre went behind a row of shops to shed his uniform.

He noticed more details about the villages. Many of the houses had sheet metal walls or roofs, which Anh explained one would not see in a regular village. "These are very poor people as they are all refugees from the war. They use what they can scrounge." Occasionally, Anh stopped and chatted with people. "My Jarai is not very good, but I'm getting better."

DeAndre was a people person. He noticed, as he had in the villages, many of the toddlers wore no diapers or other clothing below the waist. "Mom or grandma just hold them up under the arms and let them pee when they wish," grinned Anh.

Some of the older women had a weathered look about them, as though they had suffered much. All of them had darker skin than the Kinh. Most, but not all, wore traditional Jarai clothes which consisted of a modest black dress decorated with multi-colored complex stripes, often with black headbands decorated the same as the dresses.

Some women were unencumbered by clothes above the waist. A few wore rumpled western clothes, but DeAndre wondered if it was because those were the only clothes the displaced people could find.

In either case, the women were slender and shapely. Most were intrigued by DeAndre as they had never seen a black man before, some even daring to demurely come closer to look, though they were careful not to seem rude.

One tall young woman came closer than the others, looking intently at DeAndre, but he thought her eyes showed more of a "come hither" attitude than plain curiosity. He allowed himself a longish glance back at her, with the slightest hint of a smile on his lips. Anh noticed the silent exchange but only smiled and said nothing.

The women contrasted sharply with the men who wore only a loincloth or western style clothes.

As they approached the church, DeAndre noticed a white man talking with some of the Jarai men in the neighborhood. Anh whispered, "You must meet this man." Discreetly waiting nearby, they waited until the men left, then let their presence be known only after the men had left.

"Excuse me, ong Smithy" said Anh, making the word ong sound like ohm. "I would like to introduce an American to you."

With that, DeAndre and Taylor "Smithy" Smith shook hands, exchanging pleasant smiles and words as they spoke.

"Smithy is a missionary, DeAndre, and has been here for many years."

"Very good to meet you, Smithy. I'm DeAndre Washington, an advisor to the Rangers here in Kon Tum. Anh has been kind enough to show me around the Jarai area."

"Oh, he has, has he?" chuckled Smithy. "You know, of course, that he is a rebellious young man."

Anh laughed at that and it was obvious that the two knew each other well enough to share a little good-natured teasing.

"So, why did you decide to visit the Jarai?" asked Smithy.

"I dunno exactly. I'd see them out in the field, and we went through a few villages. Just got curious, I guess."

"As you can see, they are quite different from the ethnic Vietnamese."

"Yeah. I've noticed the dissimilarity between the Kinh and the Jarai. What gets me is how much the Kinh seem to hate the Jarai. I understand the word 'moi' means savage."

With that, Smithy grinned. "Methinks, you know more than you say. Using words like Kinh and moi gives you away."

"No. Not really. I just heard the words bandied about and Anh helped me out. But Smithy—can you tell me why the Kinh seem to hate them so much?"

"The answer is something you can identify with. As a black man, you understand prejudice. It really is as simple as that. Besides, the Kinh want the land for tea and coffee plantations," replied Smithy.

The conversation continued for a few more minutes as DeAndre learned Smithy had lived in Vietnam for fourteen years, spoke fluent Jarai, and had translated much of the New Testament into Jarai. He realized Smithy could teach him a lot, including the language.

Soon, Anh mentioned that the two had been out in the open for quite a while and that might give a Ranger a chance of seeing DeAndre in civilian clothes in the Jarai area. Both DeAndre and Smithy understood. Though Smithy offered to take the conversation inside the church building, DeAndre declined but also told the missionary he would really like to get to know him better.

The two parted with comments about getting together again, but no definite plans. With that, DeAndre returned to the place behind the shops and put his uniform back on.

CHAPTER SEVEN

The captain had no more than walked back into the hootch when Sergeant Rocatta ran in and breathlessly said, "All hell has broken loose, Captain. Thiếu tá Hoang wants you at headquarters now!"

As DeAndre double-timed to the head shed, he pretty much knew what he would hear—that the Communists had finally launched an assault in their area.

Thiếu tá F gave his advisor a scowling, "Where the hell have you been" look when Captain Washington entered the room and plopped himself down next to Anh.

The Ranger intelligence officer continued his briefing. On April 4, the Communists had indeed launched a major attack in their area. The enemy's invasion across the DMZ had been amazingly successful. It was more like World War II than previous fights in Vietnam, including the Communist use of modern Soviet tanks.

As they overran the big ARVN artillery base at Camp Carroll and captured the city of Quang Tri, they were well on their way to Hue. Both the ARVN and the Americans were surprised at the size and power of the invasion. The intelligence officer continued by describing yet another major invasion, this one far to the south, to the northwest of Saigon, and obviously intended to take the capital city.

They had already overrun Loc Ninh and were threatening the city of An Loc. DeAndre's memory flashed back to his time at the old American base camp near An Loc when he was with the 1st Cavalry Division.

Now their turn had come. The Communists were coming across from Cambodia in an attempt to split South Vietnam in half by taking Kon Tum and Pleiku, then going all the way to the coast along Route 19 and past An Khe, another former home of the 1st Cavalry Division.

The intelligence officer outlined the buildup of enemy forces across the border. They had built up a huge invasion force of three divisions, which, like Quang Tri and the DMZ, was augmented by Soviet made tanks. Arrayed against them were two ARVN divisions with only two armored cavalry squadrons equipped with small, light M-41 tanks, which had little chance against the larger T-54 Soviet tanks.

Also augmenting the friendly forces were elements of the Vietnamese Airborne and Rangers. There were still some American Cobra gunships at Pleiku. There were units of the ineffectual South Vietnamese air force available, but there were only about a hundred forty American jets, some flying out of Thailand. Of course, the huge B-52 bombers out of Guam were available, but they couldn't be used for close support.

The situation looked grim.

DeAndre thought he knew why the northerners were attacking now. First, there were no more American ground combat units in the Central Highlands. Second, the lingering monsoon weather caused low clouds which required all-weather planes. Third, the Communists thought the southerners would be too wimpy to fight after the Americans had left. Fourth, the Americans had so fixated on the Tet Offensive of 1968 that they acted as though every major action would take place at Tet. When the Lunar New Year came and went in 1972, everyone relaxed—except the Communists.

The northerners first made a diversionary attack in the coastal area of Binh Dinh Province in an attempt to con the ARVN into moving troops out of the Highlands. It didn't work.

The first attack near Kon Tum was on Firebase Delta. It was almost overrun, but only a third of it was captured. Some of the American Cobra gunship helicopters from Pleiku did a job on the invaders and they eventually withdrew.

But Delta was just the first of the firebases along Route 14 and Rocket Ridge. Next were bigger bases at Dak To and Tan Canh, defended by the ARVN 22nd Division. The initial Communist attack was on Tan Canh, led by a column of T-54 tanks. The advancing armor spooked the ARVN troops who withdrew using a hasty unplanned retreat. Northern troops poured into the hole while the ARVN command bunker was shorn of its radio antennae, leaving the base incapable of communicating with any other support. The American advisors fled to a nearby hill where they set up their radios and tried to bring in close air support, but the foggy weather made any airstrikes impossible.

Dak To fared no better. Again, the light ARVN tanks

were no match for the T-54s, one of which got close enough to fire directly at the command bunker. Between the intensity of the artillery barrage and the large number of northern infantry troops, the ARVN were overwhelmed. The 22nd Division ceased to be an effective fighting unit.

Some of the southerners attempted to cross the Dak Poko River, but without any cover, were cut down by the enemy. All told, the Communists captured 23 105mm and seven 155mm howitzers, plus large amounts of ammunition. It now became obvious that the firebases all along Rocket Ridge would have to be beefed up, even though they were all totally dependent on helicopter support. The way was open for the attackers who were only 25 miles from Kon Tum.

Then the operations officer stood and said that the 11th Ranger Battalion's mission was simple. They would be helicoptered to the old Polei Kleng Special Forces Camp to augment the ARVN battalion already there. Though not part of the Rocket Ridge group of firebases, it did sit on the western approach to Kon Tum. Movement would be by Chinook helicopter—the big twin rotor craft normally used to supply troops in safe areas. He instructed the company commanders to be ready on the landing zone at 0800 the next morning.

That didn't give much time to prepare, especially for logistics. As much food and supplies would have to be taken on the helicopter, and the S4 logistics officer briefed the company commanders about what they needed to bring with them. It couldn't be everything as the helicopters could take only so much weight.

Thiếu tá Hoang finished the briefing. "Gentlemen, we all know this is on very short notice, but we also know this can't be helped. I will tell you that I know the commander of the ARVN battalion we will be working with. He's both a good man and very competent. We will be each other's shadow, but ultimately, he will have command as he has control of the ARVN artillery. I will handle any gunship or close air support. Any questions?"

Unbeknownst to the attendees of the briefing, events were changing once again. During the past three days, the Communist forces had greatly increased their artillery bombardment of the Polei Kleng camp. Five hundred rounds a day were intended to systematically destroy the fighting bunkers on the perimeter—and it was successful. Despite the pounding from numerous B-52 strikes, the enemy kept coming.

The morning of Thiếu tá Hoang's operations meeting, the Communist attacked with a column of tanks and infantry. Having no anti-tank weapons, the ARVN knew they were doomed. In as orderly a fashion as possible, they abandoned the Polei Kleng camp. Though no longer capable of fighting, the battalion's officers and sergeants did a good job of leading the remaining troops over the mountains to Kon Tum. If the Rangers had been sent to rescue the Polei Kleng camp, they would have been a day late and a piaster short.

Rangers are light assault troops, and neither trained nor experienced in the kind of static defenses such as they were about to do at Kon Tum. The sergeants and men frantically built more fighting bunkers to bolster the perimeter and provide room for them beyond the ones already built for the regular ARVN troops. Nonetheless, Kon Tum was the provincial capital and gateway to Pleiku, which was the terminus for Route 19, the major road leading to the coastal city of Quy Nhon. If Kon Tum fell, most likely the Communists would split the country in half.

As the Polei Kleng camp survivors straggled into Kon Tum, Thiếu tá Hoang was surprised to see them in good spirits and spoiling for a fight to avenge their fallen comrades. He talked with the ARVN battalion commander and the two of them decided to use them as a ready reaction force within the perimeter and keep the unit together.

In the next few days, DeAndre spent much of his time in the operations center with Anh, plotting the advancing Communist forces as well as talking with Colonel Arredondo. Not wanting to take himself out of the situation, he asked Anh for a favor: go into the Jarai area of Kon Tum and tell Smithy to get back to Pleiku immediately, before the Communists got to Kon Tum. DeAndre knew that the two Christian missionaries would be toast if caught by the Communists.

Anh returned and said Smithy would not leave. If his Jarai parishioners had to stay, he'd stay with them. DeAndre didn't know if Smithy was deeply faithful or blindly stupid.

A small convoy of trucks arrived at the Ranger camp,

loaded with Light Anti-tank Weapons, known to all as the LAW, and meant for the Rangers. Though lightweight and easily carried (some American soldiers referred to the LAW as the toilet paper tube) the LAW was effective enough to punch holes in a T-54 tank. Also in the shipment were a few TOW missiles. TOW stood for Tube-launched, Optically tracked, Wire-guided. It was a very modern weapon, having come into service in 1970.

But it was also much larger than the LAW and a multi-person crew was needed. It also had a much bigger punch than the LAW. A platoon of Rangers had previously been trained in their use, and those crews took over the missiles, putting them on the western and northern parts of the perimeter, under the control of Đại úy Pham.

As all experienced combat leaders know, if you want to hear the gods of war laugh, just make plans. That's what happened to the 11th Ranger Battalion. As soon as Thiếu tá Hoang said his troops were ready, new intelligence came in. The Communists had turned away from their attack to the west of Kon Tum and instead moved towards Dak To. Plans had to be made anew to head towards Dak To.

But the respite soon ended. The supposed attack on Dak To was a feint. The Rangers returned to Kon Tum just before the enemy sent wave after wave of assault troops at them. The Rangers and other ARVN troops fought valiantly, mowing down the northern troops with machine gun fire, Claymore mines, and accurate M-16 fire. Thiếu tá Hoang's artillery observer directed the two batteries of 105 mm artillery and worked closely with DeAndre who brought in close air support. It was important there be no aircraft in the area while the artillery fire came in, so the two worked next to each other as they brought steel hell down on the attackers.

Using the giant B-52 bomber, the American Air Force unleashed arc lite missions in the enemy's rear areas, wrecking their supply trains.

But the B-52s didn't get all the T-54 tanks. DeAndre was surprised to see them deployed in perfect coordination with enemy infantry. The Russians had done a good job training the commanders. Under cover of darkness, the Communists had amassed four entire regiments of infantry and twenty-four T-54 tanks—a formidable force indeed. Their intelligence was good too—their point of attack was a regular ARVN unit, not the Rangers, and the perimeter was indeed breached. Leaving his forward observer and DeAndre to continue coordinating the supporting fire, Thiếu tá Hoang personally led two companies, reinforced by two of the jeep mounted TOWs, and counterattacked.

To everyone's surprise, after DeAndre told the helicopter units based in Pleiku that the enemy had tanks, four Cobra gunships showed up, all outfitted with TOW anti-tank missiles. He didn't quite know how to use them as he'd never even heard of a helicopter-mounted TOW before, but the pilots came up on his radio frequency.

In the garbled tone distinct to helicopters, the pilots' voices sounded as if they had taken hold of their cheeks and were flapping them rapidly against their teeth. "Black Flag 5 Alpha. Ahhhh, no problem, Dusty Acres 6. We can do this without a FAC," said the leader of the gunships, referring to a forward air controller. They came in at high altitude where they could recon their own targets as they stayed above the arc of the artillery fire. Once they spotted a target—or targets—they let DeAndre know, he would tell them when the tubes were cold and the birds went in for the kill. Combined with the jeep-mounted TOWs of the Rangers, they did a job on the T-54s.

But on came wave after wave of NVA troops. DeAndre realized his own M-16 was needed. Telling SGT Rocatta to listen for anything that required his attention on the radio, he moved forward and joined a Ranger squad, all in the prone position and taking careful, well-aimed shots at the advancing enemy. On the one hand, he admired their discipline in making sure each shot counted, yet also realized that the enemy still advanced, though with noticeably fewer people.

For the first time, he heard the dreaded sound in this fight of a .51 caliber machine gun. Adding to the din of AK-47s, RPD machine guns, M-16s, M-60 machine guns and the occasional exploding grenade, the big .51 brought a lot of firepower to the fight. The Ranger to his immediate right screamed as his steel helmet flew through the air, revealing a long bloody wound on his scalp. Judging from the size of the hole in the helmet, DeAndre figured it probably came from the .51. Lucky man that the bullet wasn't an inch lower.

An enemy bullet hit right in front of DeAndre, kicking up dirt that temporarily blinded him. He spotted movement to his left and shot an NVA soldier about to throw a grenade in their direction.

Thiếu tá Hoang moved the artillery fire to counter the threat coming from the NVA in front of DeAndre. As the 105mm rounds began exploding among the enemy, the captain began to notice a slump in the NVA's fire. The pounding continued and had great effect.

The Ranger company commander passed the word—on his whistle signal—the Rangers would counterattack the NVA just as the artillery barrage lifted. As the Rangers moved forward, DeAndre dragged the severely wounded Ranger to the rear, hoping to find an aid station. Two Rangers appeared with a litter, and they took over the rescue from the American.

DeAndre dashed back to his own foxhole. To his horror, he found SGT Rocatta severely wounded, a bullet having hit him right on the left shoulder joint. He was conscious, but just barely. Trying to avoid causing any more pain, DeAndre took the radio off Rocatta's back, then tore off his shirt in order to see the wound.

Besides seeing the horrible damage the AK47 bullet had done to the shoulder, DeAndre also saw the two tattoos on Rocatta's chest—and they were very disturbing. Setting aside his disgust, he quickly came to his senses and did his duty.

Calling SFC Wozniak, DeAndre explained what had happened to Rocatta. Within a few minutes, Wozniak showed up in a three-quarter ton truck. After loading him on the truck, Wozniak stayed with Rocatta while the truck drove to the helipad where a Dust-Off medical helicopter took him to the 67th Evacuation Hospital in Pleiku. Both Wozniak and Washington knew their young sergeant would survive but had a very long recovery road ahead of him.

Thanks to copious aerial resupply during the fight, Thiếu tá Hoang was able to rearm and re-provision his troops quickly and set off in pursuit of the retreating enemy. After a day, Hoang realized his exhausted troops may have wanted to fight, but they were just too damned tired. The battle was over. Kon Tum was saved. The enemy's 1972 Easter-tide campaign to end the war had failed, leaving over 4,000 NVA dead near Kon Tum alone.

And Deandre had his answer—Thiếu tá Hoang was indeed a fierce and skilled leader of combat troops. He'd been excellent, and at least for the time being, DeAndre forgot about Hoang's hatred of the mountain tribes.

The Rangers staggered back to their camp, counting their losses, but nonetheless knowing they were victorious. After asking Thiếu tá Hoang if he could help in any way, he was merely asked to walk a sector of the perimeter and be sure the camp was properly secure. He found things in good shape and all DeAndre had to do was inform Thiếu tá Hoang that one particular bunker needed more ammunition. The order was given that all personnel would sleep in or around their assigned fighting bunkers, with one awake and two sleeping.

CHAPTER EIGHT

Neither of the two remaining Americans gave a damn about being dirty and sweaty. They flopped down on their bunks, needing sleep. But DeAndre had too much adrenaline in him to fall asleep immediately. More adrenaline. More addiction to risky behavior. After an hour or two, his body finally drifted off into a long and hard sleep.

He dreamed of normalcy. It helped him balance the hate and death he'd seen that day. His mother, cooking a favorite meal. A stateside assignment as a garrison soldier. A meal of comfort food at a diner. An erotic dream of a nubile Susan in the shower with him.

After the image of Susan's gorgeous body faded, he was dreamless. The sun was well into the sky when he entered that nebulous space between sleep and wakefulness. His thoughts again drifted to Susan, and how she once asked a simple question. "DeAndre, you're a pretty laid-back dude. Why in the world did you pick being an infantry officer as your career? You're not prone to violence. I've never even heard you get angry or yell."

Just before his eyes opened, DeAndre thought about that question. Why had he indeed? He knew the answer, of course, but now the day had begun. A brisk shower woke him up, then a change into clean fatigues and he joined SFC Wozniak in eating LRRPs for breakfast.

For a while, there was little for him to do. Kon Tum and the Ranger base camp were in bad shape and needed to be rebuilt. The mess hall had been destroyed so it was LRRP and C-rations all the time. Numerous sections of the city had suffered as homes were blown up by artillery fire, markets ruined, transportation as simple as bicycles had been damaged.

DeAndre wasn't needed to advise the Vietnamese how to rebuild the military areas, so he drifted back into the Jarai area to see if there was anything he could do. He found Smithy looking tired as he surveyed the heavily damaged church building.

"Ah, welcome back, DeAndre. I'm sure you have been quite busy lately. But really—thank you for helping keep the Communists from taking over. There would have been ever so much more bloodshed."

After chatting amicably for a few minutes, DeAndre ended the small talk by saying "Okay, Smithy. What can I do here?"

Tossing a shovel in DeAndre's direction, Smithy grinned, saying, "Join us as we get rid of the rubble that was once our church. Many of our people have their own homes to rebuild and can't help here."

With that, DeAndre's shirt came off and soon the sweat was pouring as he scooped debris into an old ox cart. Some of the Jarai walked by on the way to find supplies were startled a bit at the sight of the fit body of a black man. The women were a little more curious and again, as had happened during one of his early visits, the same tall woman walked by, slowed down and openly cast an approving look his way.

"Ah, that is Chel," said Smithy, after she left. "She's not one of our flock, but I talk to her often."

And DeAndre got a lesson in the Jarai's culture. "It's strange to us westerners because among the Montagnards, marriages are arranged. But it is common for the woman to select the man then tell her family and have the union "arranged." Also, the family tree is from the mother's side. After a wedding, it is the man who moves in with the bride's family. The tribes intermarry a lot, so while they may not have seen a black man before, having the women check out a man who is not Jarai is no big deal.

"None of the men will say anything to them about gawking at a black man, but Chel is obviously both curious and attracted to you."

Equally startled was Smithy's wife, Lee Anne, when she came around the corner towards the church and saw DeAndre shoveling. After Smithy introduced them, she grinned, saying, "It's not every day I see a shirtless American black man here in the mountains of Vietnam."

Turning to Smithy, they talked about the various things that needed to be done at the church building, as well as the small hospital run by American Dr. Patricia Smith. Almost unbelievably, Smithy and Lee Anne's house had been spared.

DeAndre wasn't too worried about being spotted by a Ranger. They were all occupied repairing the camp's defenses and housing.

"Hey, I know you're a tough guy, DeAndre, but let's take a break and get some water in us."

"Sounds good to me. When you gonna hold services again, Smithy?"

"Oh, this Sunday. We don't need a building to worship God."

"Wow. That's different. Americans can't imagine not having a church building for Sunday service."

That made Smithy smile. "Oh, we have other things to do than criticize our American brethren. It also sounds like you are a regular at your church."

"I am when I'm visiting my mom. She has been a regular at her old church since before I can remember. She took me on the last day I was home and the whole congregation joined in laying on hands in prayer."

"Tell you what—I realize you don't speak Jarai, but I bet you will know what we're doing. Why not join us this Sunday at 7. We'll meet in the small thicket of trees back there," Smithy said as he indicated with his finger where they'd meet.

Picking up his shovel again, DeAndre said he looked forward to being there.

Then Chel ambled by again, unabashedly checking him out. His memory flashed quickly to Susan, the other forbidden fruit in his life. She also talked to Lee Ann.

With the reconstruction of Kon Tum, and no threat of enemy attack, DeAndre had a lot of time to spend in the Jarai district. Anh had to remind him again to be careful he wasn't seen by any of Thiếu tá Hoang's people, so he took to being inside the church to help with some of the carpentry work. Overhearing Smithy talking to Lee Anne one day about not having any money to buy some lumber, DeAndre dipped in his pocket and handed them some piasters to buy it with.

"Wow! Thank you, DeAndre! We truly appreciate it. What can I do in return?"

Pausing for a moment, he answered "Teach me Jarai."

"Done" said Smithy, but it also set him to wondering what was going on with his new American friend. When DeAndre had pounded a few more nails, Smithy asked him to stop and join him for a break. They wandered outside to an area near the church that was secluded enough to provide privacy from the Rangers.

"Just curious, DeAndre. Curious as to why your interest in the Jarai."

"I'm not sure, Smithy, but I think it has something to do with what I've seen and heard about the Kinh's treatment of the mountain tribes. Yeah, yeah, I know—as an American black man, I've had my own brushes with discrimination, but nothing like what these people endure. When we go out on the so-called "recon missions" with the Rangers, it takes all I have to not get pissed at Thiếu tá Hoang, tell him to knock it off, and get the damned Kinh the hell out of the mountains."

DeAndre took a moment to listen to himself and what he'd said—and how he'd said it. He recognized the indignation in his voice and noted he had gotten louder the longer he talked. He realized he had just revealed a deeper part of himself to Smithy.

Smithy took a long pause, long enough to make De-

Andre regret he'd spoken so openly. But when he started talking again, it was calming.

"DeAndre—I can see you are a passionate man. But please be careful. I agree with you very much, but we are fighting both a strong cultural pull and official government policy. If your Ranger boss catches wind of how you feel, there might be greater consequences than merely being withdrawn to Pleiku."

He paused, wondering if DeAndre had any questions. He did not, so Smithy continued, "I have seen Hoang kill Jarai, and I mean, he himself killed them, not merely ordering some poor soldier to do it. I was in one village when his unit made a visit, but I stayed out of sight. When the Rangers kicked an old woman out of her house and kept on kicking her, the son came out of the house screaming and tried to stab one of the Rangers. When Hoang saw that, he ordered his men to hold the son's arms behind his back, then he literally sliced the man's head off with a machete. It took three hacks before he was successful. DeAndre, I almost puked right then and there."

Another pause, even longer than before. Smithy asked, "Have you ever heard of FULRO? The acronym stands for the French words for the United Front for the Liberation of Oppressed Races."

"No."

"It started back in the early sixties, but in1964 some of the tribesmen who had been trained and armed by American Special Forces troops staged a rebellion in Ban Me Thuot. Sure, the ARVN sent in an entire division to put them down, but that didn't end the movement. I find it ironic that the Americans trained these people to fight for South Vietnam, which they did and did well, yet the Kinh's hatred for them remained strong."

"Yeah, I can relate to that. I've often wondered why American black men joined the US Army during World War II, to fight for a country that still wouldn't even allow them to eat at the same restaurant as white people."

"Some of the younger Jarai around here have joined FULRO. Even some of the American Christian missionaries have helped them by smuggling in arms."

"Anybody I know?" smiled DeAndre, casting a raised eyebrow at Smithy.

"No, not me, but I have no quarrel with the idea, except that it brings violence, and Jesus eschewed fighting. He went to His death willingly, with no resistance."

"Why are you telling me this?"

"To give you options you may not have known existed," replied Smithy. "There are some things to keep in mind. First, the tribes get along very well with each other. Of course, they identify with their own language and traditions, but they intermarry with no problems. More and more, they think of themselves as Hill People—Montagnards—rather than Jarai, or Rhade or Katu or whatever. The fancy academic term is ethnonationalism. They themselves have started using the term 'Dega.'"

"Just sayin', that's all. Just sayin'. How would I get in touch with someone in FULRO?"

"Well, she's been sizing you up. Chel is part of them, not as a fighter, but as someone who is on the logistical side of things."

"One big problem, Smithy—I can't speak Jarai, and it doesn't sound as though she can speak English."

Smithy put on a big toothy grin. "Oh, ye of little faith. You asked me to teach you Jarai. We can do that. Then you can practice Jarai with Chel."

Anh came around the corner. "Captain—maybe you've stayed here too long. I suggest you get back to the compound."

"Why? Has Thiếu tá Hoang said something?"

"Yes, he asked for you a few hours ago and Sergeant Wozniak went up to the headquarters. Hoang is not pissed, but he wants to know why you didn't come."

"Ooooo, okay. Thanks for the heads up, Anh. I'll be thinking of some excuse for not being there."

After saying goodbye to Smithy, DeAndre found his usual spot, changed back into his uniform, then walked back into the camp with Anh. "I think I'll just tell him I was hanging out with some American helicopter pilots, drinking beer," said DeAndre.

"Sounds good, but I suggest you have a beer before going over to headquarters, so you at least smell like you were drinking," said Anh with a grin. With that, he broke away so as not to be seen with the captain.

CHAPTER NINE

Reporting to Thiếu tá Hoang in his office, he was told that most of the reconstruction work on the camp was done and it was time to resume operations. The replacement troops needed some on the job training, and he would explain all that at a briefing in a few days. If he had any suspicions as to DeAndre's whereabouts, he didn't show it.

Sure enough, at a briefing three days later, Hoang announced the Rangers would resume rousting out the nearby Jarai villages. DeAndre also noticed a South Vietnamese general seated up front next to Thiếu tá Hoang. While saying nothing, he smiled his approval of the plans for the next "reconnaissance" mission.

"Who's that?" whispered DeAndre to Anh. "Later," was the barely audible reply.

At the end of the staff briefing, the general stood up and made a few comments. Anh sniffed as he translated for DeAndre. "He approves of the mission of the next operation because it fits right in with the overall mission of subjugating the moi and for the Kinh to take over the highlands." DeAndre noticed that the general looked directly at him before saying something else. Anh merely said that he would tell DeAndre later.

Walking back to the hootch, DeAndre asked, "What the hell was that all about?"

"General Ngô Du is not just a general. He is also the government's person in charge of II Corps. What he says goes—period. He reports directly to President Nguyen Van Thieu. It's hard to say which one is the most corrupt."

"Are you saying that these "recon" missions are approved by Ngô?"

"Approved? Hell, he ordered them. Most South Vietnamese units in Two Corps do similar missions."

= = = = = =

DeAndre started his Jarai lessons the day before the first operation began. Smithy and Lee Anne worked him hard for four hours, and he was an adept student. He told them both he would be gone for a few days because of the Rangers' operations, but he didn't know exactly how long. Just before he left, Chel joined the group.

"You're learning Jarai and she is learning English," remarked Lee Anne. "And we are happy to help her learn. She is a very bright woman." With that, Chel beamed. It was obvious her English skills were good enough to understand the compliment.

Captain Washington grew weary of advising his Vietnamese counterpart. Hoang was a rich man, made wealthy by moving Kinh out of the Jarai villages and into larger Vietnamese villages, in total contravention to the official government policy of assimilation.

The bribes he took entitled the Kinh civilians to both

a military escort and Army trucks with troops to move their possessions. Washington had watched Thiếu tá Hoang persecute the Jarai and allow his troops to steal from them. Other sources of income for the major might be the red envelopes from rich parents given in the hope of keeping their soldier sons out of the field and punching a typewriter.

DeAndre suspected he was also black-marketing government supplies. Hoang might also have some ghost soldiers on the battalion payroll. DeAndre couldn't prove any of these allegations, but he enlisted Anh's help, asking him to do a bit of amateur spying on the staff officers. Under the pretext of wanting to read the operations orders, he had Anh with him, but he didn't do it so often as to bring attention.

The next "reconnaissance" mission on a Jarai village was worse than any DeAndre had seen before. The village had been gutted during the Communist attack. Most of the long houses had been burned to the ground, animals dead or scattered. The villagers were sleeping on the ground, caring for their children by erecting little tents made from military ponchos. The few possessions they still had were stashed in bags and boxes and kept on the ground next to them. Cooking fires were next to each family. DeAndre could see that what little meat they had came from the jungle. Though the farming fields were relatively undamaged, there were no vegetables as it was still spring.

The Rangers had just spent a prolonged time in combat and still had a certain amount of bloodlust. Rather than merely pushing people, they now punched them and dragged them. One soldier was seen mounting a bayonet on his M-16, but that was too much for the sergeant who stopped the stabbing. There were no Kinh in the village. They had evacuated before the fighting began. Thiếu tá Hoang didn't get any red envelopes, but he oversaw the Rangers steal what few precious possessions the Jarai had.

DeAndre looked up when he heard very loud shouting in Jarai and Vietnamese. Two Rangers came out of a shallow bunker roughly manhandling a young Jarai man, all three screaming. Then a sergeant appeared coming out of the bunker, holding a Chinese made AK-47 rifle. DeAndre didn't need Anh's translations to understand that the Rangers had just captured an armed FULRO fighter. Bringing the prisoner in front of Hoang, there was a sharp exchange among the Vietnamese. The rifle was left with Hoang's driver for safekeeping.

Three soldiers then shoved their victim into a small bamboo thicket. Three shots rang out and the three Rangers reappeared. South Vietnamese justice had been dispensed. DeAndre noted the presence of General Ngô, whose smile and nodding head indicated he approved of the entire episode.

As far as DeAndre was concerned, this was enough to make a formal report to his boss, Lieutenant Colonel Arredondo. He composed it carefully after returning to Kon Tum. Still pissed at what he'd seen, he included some of his suspicions as well and pondered why General Ngô Du was on the operation. After showing it to SFC Wozniak, who had seen most of the incident, and getting Wozniak's nod, he got on the radio and requested a helicopter pick him up for an appointment with Arredondo. He got it. The next morning at nine, he was at the American helipad, then made the short hop to Pleiku.

"Good timing, Captain. I need to see you too. But you first," said Arredondo after the two had exchanged salutes and sat down.

Reading the report slowly and carefully, DeAndre then put it slowly on his desk. It was quite a while before the colonel said anything.

"Washington, both of us are soldiers and both of us want to do our duty, but something tells me that you've had it with the South Vietnamese, just as much as I have. My only real problem with the US pulling out of Vietnam is that we didn't do it sooner."

This kind of openness rocked DeAndre. He'd never had a superior officer confide in him before. A long pregnant silence ensued.

"But on to other news, my good captain. In July, all battalion level advisor positions will be eliminated. The American pull out continues. That means you'll move here as my replacement."

Breaking into a grin, he continued, "Don't get your hopes up for a promotion. The job will only last for a few months until this one too is eliminated. I have no idea where you'll go after that. As for me, I'm going back to the world."

"Damn, colonel! Not that I particularly crave your job, but I also have to ask what I will do in your job. With no battalion advisors, there won't be anything to do."

"Right you are, my good man. Just cozy up to the

province chief and General Ngô, and that's about it. But, while still in your current job, you are to instruct Thiếu tá Hoang in calling in airstrikes and gunships. I know he has an artillery forward observer already."

"Sure. No problem doing that."

With that, LTC Arredondo went to a cupboard and took out a bottle of very good single malt Scotch and two glasses. "Shall I say it? It's five o'clock somewhere. Let's relax a bit, eh? Limit of two."

Cradling his glass, DeAndre asked, "What about Sergeant Wozniak?"

"Not to worry. His orders will be coming down in a few days. He's been in-country long enough to earn an early out. He'll be on that big silver freedom bird very soon."

Returning to Kon Tum, he sat down with SFC Wozniak and gave him the news about both of them. Out came his own hidden bottle of good Scotch.

"I can't say I envy you, Sir. On the one hand, you won't have much to do, but you will be pretty much at the mercy of the ARVN. I hope the aviation units stay in Pleiku at least as long as you're there."

The two men had strongly bonded during their time together, and though both were happy to be out of combat, both also knew the camaraderie that is a large part of military life was at an end. They both toasted SGT Rocatta and hoped he was recovering well back at the huge Valley Forge Army Hospital, just outside of Philadelphia.

Amazingly, Wozniak's orders came the very next day.

"Hey Captain—I'm going to the Infantry School at Fort Benning."

"Ah yes—the Benning School for Boys," teased Washington. "But I wouldn't think the place will be all that busy. They've already closed the Officer Candidate School and with the draft almost gone, I'd be surprised if they still have Basic Training there. And I don't know if you are mean enough to be a Black Hat at jump school."

"I don't really care, Sir. Benning has one of the best Senior NCO clubs in the Army."

The two joked and laughed about Fort Benning—the Home of the Infantry—which continued after DeAndre poured another scotch. They got a little raucous, but nothing objectionable. As the conversation cooled off, the pauses grew longer.

Finally, Wozniak said, "Sir, there's something I need to tell you." Though that made DeAndre nervous, the sergeant went on, "It's about Sergeant Rocatta, sir."

"Oh great! You've heard from the kid? How's he doing?"

"No sir. I haven't heard from him." The overuse of the word "Sir" made DeAndre nervous all over again.

"On the day he was hit during the attack on Kon Tum, you called me to evacuate him. By the time I got there, you'd already bandaged the wound, but he was in pain. Big time pain. The two ARVN medics gave him some pain killers, then they helped me load him in a truck. He was pretty hopped up on pain meds."

SFC Wozniak paused, took a deep breath, and went

on. "He said, 'Well, at least I won't have to work for that damned nigger captain anymore.'"

This was no surprise to DeAndre as he'd seen the tattoos on Rocatta's chest. 'So you see, my good sergeant, that he was no moody teenager—he was a virulent racist."

"All the more to your credit, Captain."

"Naw. You know this as well as I do… that soldiers in combat pride themselves on the idea that when the bullets flew by, the only color is olive green. I just did what I could for another green person."

The light-hearted approach both surprised and pleased Wozniak.

Two Scotches later, DeAndre took a noon-time nap in accordance with good Vietnamese custom, then went over to the Ranger headquarters to arrange a time to train Thiếu tá Hoang in calling in air strikes and helicopter support.

While still sitting in the major's office, DeAndre informed him that both he and SFC Wozniak would be leaving, though he would only be going to Pleiku. This didn't surprise Hoang as he knew the Americans were leaving. But he was very happy to hear about being trained to call in close air support and helicopter gunships.

DeAndre found an eager student as the Ranger officer saw this as a big advancement of his power. DeAndre arranged for one gunship out of Pleiku to be used in the training, but it was not live fire. Hoang, being an intelligent man, soon picked up the skill. DeAndre could not arrange for a FAC to partake in air strike training, but DeAndre wrote out a detailed list of things to do, including several if/then situations that called for decisions. He had Anh translate it. That seemed to satisfy Thiếu tá Hoang who was a bit disappointed not being able to call in strikes using American aircraft.

Three days after the training, DeAndre was at headquarters when Anh told him Hoang did in fact try his new-found skills. Of course, he practiced them on some Jarai by calling in gunships on a small band of them in a jungle clearing, claiming they were FULRO. That put DeAndre over the top. He left the headquarters building, going to his hootch. There, total rage overtook him. "He kills people just because they make good targets? What total bullshit!" he yelled, swinging his fists in the air. "No fucking way the US should be supporting such barbarity. Screw South Vietnam—let the Commies have it. These mountains belong to the 'Yards—not the damned Kinh."

DeAndre began to look around for something to throw in his truly purple rage. "Shit, I'd rather fight with the tribes than the damned South Vietnamese government."

Wozniak walked in. "Sir, you might want to keep it down just a bit. There are some people getting curious at your screaming and yelling. I'm afraid one of them will speak English," he said, implying dire consequences if the Vietnamese found out.

DeAndre whirled around to see Wozniak, quite calm and reasonable. "Yes sir, I just heard what the bastard did. I am damned glad I'm going home soon. I'm like you—I can't handle Hoang's unbridled loathing of the 'Yards."

"Thanks, Woz," using a familiarity he hadn't used before with the senior sergeant. "That was stupid of me."

"Not to worry, Sir."

DeAndre realized his heart was racing. He looked in his shaving mirror and saw how red his face was. He needed to calm down.

The two sat down and said nothing to each other. And it was good. They shared a beer, though unfortunately, the only beer available was Vietnamese Bier 33—known to all as "Ba Mui Ba." Neither the captain nor the sergeant particularly enjoyed it, not so much because it didn't taste good, but merely because it was Vietnamese made.

With that, DeAndre left the military post and went into Kon Tum city to see Smithy. After reminding Smithy of his pending departure, he went on, "Smithy, though I don't totally understand your devotion to the Jarai, and I'm not as far along in my faith as you are, I still want to strongly encourage you to leave Kon Tum and go to Pleiku." He went on to explain how Hoang had practiced his skills by having ARVN helicopters strafe Jarai in a jungle clearing.

"So that's it. We wondered why the helicopters did that. It was obviously an attack of some sort, and I just kind of blew it off thinking the Vietnamese believed they were attacking FULRO."

"And that's what I'm getting at, Smithy. With no American military looking over his shoulder as he has now, Hoang is going to get bolder. I certainly would not be surprised if he intentionally goes after you simply because you support the Jarai. I implore you, Smithy—come with me to Pleiku. I understand you started a number of churches in the area."

"Let me pray about it and talk it over with Lee Anne, but for right now, I'm sorry I can't hang around to talk this morning," said Smithy. "I need to go to Ia Chim with an oxcart and bring back three of the Jarai who were wounded by Thiếu tá Hoang's helicopters the other day. I'll be taking them over to Dr. Pat's hospital."

With that, Smithy climbed aboard the primitive wagon with two Jarai men. As he pulled away, Smithy shouted over his shoulder, again inviting him to the worship service on Sunday.

CHAPTER TEN

Back in the hooch, still seething from the Hoang's helicopter attack on Jarai, he started daydreaming about Susan's question. Why indeed had he become a grunt when all he really wanted was peace, quiet, and the love of a good woman?

The answer went back to his college days. The University of Georgia had an ROTC program. All able-bodied freshman and sophomore males were required to wear a uniform on Thursday, salute the cadet officers, and attend drill in the afternoon.

True to his personality, he believed in going full-bore when doing something. He didn't want to just march aimlessly around a field with others. He wanted to be in the Honor Guard, the drill team of volunteers who wore boots with white laces, carried the old Springfield rifle that was perfect for the Queen Anne manual of twirling rifles, white gloves, and singing cadence as they marched in local parades. The khaki uniform kept pressed. The applause of the crowds. Gawd, he loved it.

The next year, he was promoted to Cadet Sergeant and made a Squad leader! This was even better. But the third and fourth years required a volunteer contract, and if successfully completed, you were sworn in as a Second Lieutenant in the Army Reserves, with a two-year obligation to serve on active duty. His juvenile attraction to being a drill team member had gotten him into a war where he was almost killed.

The one question he couldn't answer was why he stayed in the Army. He'd finished his two-year obligation and was now on indefinite active duty. His brother Dashawn had gotten out after two years. Maybe he stayed in because the Army was like a family to him. He had no wife and family. Would he get out if he married? He didn't know.

Seven in the morning seemed a little early for a worship service, but he was there. He joined the rest of the congregation in the stand of trees near the building. Though the Jarai seemed a bit startled to see an American black man join them, they all smiled in greeting. Shortly, Chel joined him.

Smithy said a few things DeAndre didn't understand but knew full well what was said when everyone turned round to look at him and bow. The service opened with singing, and to DeAndre's great surprise, they sang "A Mighty Fortress is our God," but with Jarai words.

Next was "The Old Rugged Cross," a favorite of his late father. He couldn't sing the Jarai words, but he could hum along, occasionally saying the English words in a low tone. The familiarity of the music warmed his heart greatly. Chel leaned over and said, "I teach you Jarai words. You sing too."

But he recognized more than the singing. When the people prayed, they raised their hands in the air, just like the congregation did in his mother's little white clapboard church back home. He would have to write a letter home, telling his mother and Pastor Willis of singing familiar hymns in the middle of Vietnam's mountains.

One of the men came forward from the congregation, book in hand, then after saying something to the people, read from the Bible. DeAndre didn't understand the words, but he knew he was hearing God's word.

Chel whispered, "I show you what read."

"I thought you weren't Christian."

"Almost. But must give up spirits. That hard."

When the service ended with the wonderful old spiritual "I'll Fly Away," DeAndre was transported back home. He joined in the hand clapping and even threw in an occasional "Hallelujah" for good measure. "I'll teach you English words," he told Chel, to which she beamed back an unspoken yes.

It took a long time for Smithy to disentangle himself from the congregation. Every one of them had overwhelming problems and they asked him for solutions, but most of the time, he had no idea how to help them. He knew many were leaving the worship service, but they had no home to go to. Most homes had been destroyed in the fighting. DeAndre thought to himself that it seemed to be a universal truth: those who were the least able to handle hardship were the most prone to it. It seemed like it was always the poor who suffered.

"Quick question, Smithy. I loved singing some of my favorite old songs at the service, but I'm curious as to why you use translated American music rather than something that was originally written in Jarai."

"Easy to answer. All the Jarai songs have at least some reference to the spirits, even those that are not used in their rituals. The people themselves asked to use American songs because they didn't want to be reminded of the spirits."

Lee Anne took his arm, steering him towards her home. "Okay, you two. Time to stop talking shop. DeAndre, come on over for lunch. Chel will come too. But I'm afraid we don't have much to eat," she said apologetically.

"Then let me bring some food. Do you know what LRRP rations are?"

"Oh yeah—and they're good!"

"Be right back." And with that DeAndre returned to his hootch, raided the stash of LRRPs, and brought some back to the little Smith abode, where everyone got to pick their own. Of course, Chel had no idea what any of them were, so Smithy helped her choose chicken with rice. DeAndre showed them all how to pour hot water into the bag, stir it well with the plastic spoon, let it sit for a minute or so, then enjoy.

Sitting around the table, they contentedly munched on the Tropical Bar or Fruit Bar that were also Army issue. He also noticed that Chel stayed right next to him almost all the time—and that was just fine with him.

"Okay, I have to ask, Smithy. Why do you keep Chel around me? You aren't trying to set me up, are you?"

Smithy gave him a toothy grin, saying, "No—I'm not a matchmaker. God will find just the right mate for all of us—if we let Him. God gave me Lee Anne, and it doesn't get any better than her."

"Dear Mom" began his letter. "You may not believe it, but your eldest son went to a worship service this week. Not an Army chaplain's service either, but at a native church pastored by an American missionary." He did not mention Chel, though he wanted to. He knew how much his mother wanted him to settle down.

As expected, the Army sent no replacement for Sergeant Rocatta. DeAndre wanted Wozniak to still be able to operate independently, so he asked Thiếu tá Hoang to supply him a radio man. He wanted to hump the radio himself, but he knew that would cause a huge loss of face for Hoang, so he acquiesced and he had a very pleasant young Ranger named Ngoc with him on operations. He spoke enough English to be able to communicate. Ngoc explained his name was pronounced "nop" and said that he hoped the Đại úy would be happy with his services.

The first village was not way back in the jungle. Outside of the usual pushing and shoving of old women, nothing happened to set him off. The second village that day was similar.

The first village of day two of the operation was also low key, but the fourth and last village of the operation was much tougher. There were no Kinh in any of the villages because they had all evacuated to Pleiku during the Eastertide invasion, so Thiếu tá Hoang was out among his troops. DeAndre watched as one topless young mother was torn away from her kids. When she was shoved to the ground, Hoang knelt down next to her, grabbed her hair and forced her to turn over onto her back. He grabbed one breast and twisted her nipple so hard she screamed in pain. He continued to hurt her for another minute or so, all the while DeAndre just fumed with anger.

He turned away from Ngoc and watched something else. When he heard the woman's whimpering, he turned around to look at her. Hoang walked towards him and said, "We teach moi lesson."

For some strange reason, DeAndre felt a certain coolness come over him. He didn't know why, but he did. Perhaps it was the image of such a thing happening to Chel that turned his anger icy. He wasn't sure what his mood foretold, but he had an idea where he was headed.

He still had a few weeks before going to Pleiku to live. He spent much of that time with Smithy, Lee Anne, and Chel, working on the church building, but also helping repairing some of the homes. The Jarai were surprised to see this black man laboring to help them, and he used them to practice his new-found knowledge of Jarai.

This tickled the people even more and of course such immersion helped him learn the language faster. Lee Anne was delighted to see him out in the community, working with the people.

Food relief had come into Kon Tum from areas not affected by the attack, and the Jarai were able to eat regular meals again. One day, unbeknownst to DeAndre, Lee Anne had conspired with Chel to prepare a lunch of traditional Jarai food of rice, squash, a cube or two of chicken and a little hot pepper. He wolfed down the food with obvious gusto, asking questions between mouthfuls about how she prepared the meal. Chel sat next to him, excitedly answering his questions, with only a little help with language from Smithy or Lee Anne.

The next day, Smithy stopped DeAndre before he started working to repair more Jarai homes. "You're spending almost all your time here, DeAndre. Aren't you a little worried about Major Hoang?"

"Oh, the hell with him, Smithy. I am so tired of watching that brute hurt and kill people for the pure sport of it. I'll be going to be living in Pleiku soon, and I'll be rid of him."

"Yes, and that will give him free rein to do as he pleases. Lee Anne and I have talked it over, DeAndre, and we think it best we go to Pleiku too. As you say, we know a lot of people there. It won't be like moving to a strange place. The churches in the area don't need a pastor. Most have Jarai shepherds now. I've been thinking that perhaps it would be better if I devoted what time we have left in Vietnam to try and finish translating the entire New Testament."

"Great idea, Smithy. That is truly good news."

"Would you mind if Chel came to Pleiku too?" asked Smithy with a conspiratorial look on his face. "She has family in the area."

"What? Of course I wouldn't mind, but again—are you and Lee Anne trying to set me up with her?"

"Seriously, DeAndre. You two are obviously attracted to each other. You're both fine people. She has no real reason to stay in Kon Tum. Why not give the relationship a chance and see what happens?"

"You mean something more than just being friends?"

"Yes. Of course, something beyond friends. Who knows? Perhaps you could take her home with you when you get on that big silver Freedom Bird and go home."

This was a bit mind blowing for DeAndre. Yes, he was powerfully attracted to Chel, but getting into a serious relationship with her? Take her home? That was wild. And this was all at the suggestion of a Christian missionary.

"Listen, DeAndre. You act as if missionaries are anti-relationship. Lee Anne and I are anything but that. Don't be afraid to be alone with her as if we were automatically going to disapprove. If I didn't know you, I would think you only had lustful intentions, but I know better. You've even told me that you wanted to eventually settle down. So, roll with the idea. See what happens."

"Wow! Sure, go ahead and tell her to come to Pleiku. But I need a little time to think about what you're suggesting. Oh yeah—one more thing. You told me once that she was with FULRO."

"Yes, she is. I don't really know anything more than she does some sort of logistical work. That and communications. But I suggest two things. First, you ask her to come to Pleiku. Second, you ask her about FULRO."

Lee Anne walked into the room and immediately noticed she had interrupted an important conversation. "Oh my. I'm sorry, you two. I didn't mean to intrude.

"No, no, Lee Anne. Please stay. I want to hear from you too. Smithy tells me you both are in favor of Chel going to Pleiku and you both think I should see if there might be a future with Chel and me."

"Yep. That's about the size of it on both counts. DeAndre, you both are wonderful people. Really. Why not? Give it a chance to grow. If it doesn't, nothing is lost. If it does grow, then you have a whole bunch of decisions to make."

"Okay—obviously I am very much attracted to her. And not just for her beauty. She is very smart, as well as a very caring person. I watch her work with the people who lost their homes and she will do just about anything to help people."

DeAndre hesitated a bit, as if gathering his thoughts, then went on. "She is better than me. I get very angry at the Kinh in general and Hoang in particular for the way he treats the mountain people, but I have never heard her bad mouth the Vietnamese. She is more Christian than many Christians I know."

He paused again, this time to see if the Smiths had any response, and when there were none, he asked, "But, what does she see in me? I don't get it."

Lee Anne actually giggled a little bit. "You started off by saying you are physically attracted to her, then you went on to tell of her other qualities. Okay—that's her too. You must know that you are an attractive man, DeAndre. No false modesty now—you do know that, right?"

"Yeah, I've heard that," DeAndre murmured, thinking of the comments about women being attracted to him and men admiring him.

"Also, she is tall, especially for a Jarai woman. She tow-ers over many of the men. But you—you're much taller and she likes being around a taller man. It makes her feel more feminine."

This was a new idea to DeAndre.

"But, like you, once she got past the physical attraction, she was truly astounded to see how you just pitched in to help people—people you didn't even know."

"Huh?" thought DeAndre. He hadn't helped the Jarai to impress anybody. The idea that Chel was drawn to him merely for helping the Jarai put a whole new spin on things. He let a few minutes go by as he digested all this, then finally said, "Okay, you two. I will give it a go with Chel."

Surprised and delighted at the encouragement, DeAndre said he now felt free to return Chel's attention more openly. With that, he excused himself and left to go work on another Jarai house that had been damaged during the recent fighting.

He had no idea how to bundle palm fronds to make them waterproof, so he let the Jarai themselves do that, but he did know how to wield a saw and a hammer. He'd built a crude sawhorse and was cutting some lumber to size when he felt a soft hand touch his back. Though a bit startled, his instincts also told him the touch was not hostile. When he turned, there was Chel's smiling face.

Standing up, he returned the smile with his own, showing off his pearly whites. "Good morning, Chel."

"Good morning, DeAndre" she said, returning the greeting as though she'd been saying it since she was a little girl.

Looking at the lumber and saw, she asked, "Can help?" It was a teaching moment, but DeAndre had to be careful not to sound negative, so he reached out and touched her arm, saying softly. "Can I help?" emphasizing the middle word. But Chel didn't have a delicate personality. She didn't take the correction in a negative way. She wanted to learn English. She especially wanted to learn it from her new friend. "Can I help?" as her hand covered his resting on her arm.

They exchanged smiles while he said, "Of course you can," and showed her how to hold the end of the lumber to keep it from moving on the sawhorse. She followed as he carried the cut lumber to where he was working.

"Give me two nails, please," he said.

"What nail mean?"

Reaching into the small bucket nearby, he pulled out a nail and showed it to her. "This is a nail. Give me two nails please."

He thought her hand lingered a trifle too long on his hand as she handed him the nails. As he was pounding the nails in, he thought it a good idea to return the touch. He turned, and as he said, "Thank you," he held her hand for a moment. He didn't hold it too long as he didn't want any of the Jarai men to think there was anything awry.

The two worked together for an hour or so. He taught her more, both about the language and about the work. As he taught her how to use the hammer, he both held her hand and moved closer behind her and felt her body next to his. Her reaction was to turn briefly and smile before swinging the hammer.

Returning to his hootch that evening, his head was abuzz with thoughts. Was this a truly cool thing to do? What about the Army? He had only two weeks to go before he'd be going to Pleiku, though LTC Arredondo hadn't contacted him with any specifics. Maybe Chel would move with the Smiths, and they could continue exploring the relationship then. Had Sergeant First Class Wozniak noticed he was acting differently? For that matter, did he know he was working in the Jarai area of Kon Tum?

In the morning, as they walked together to the mess hall, DeAndre decided to bring the subject up.

"I'm curious," said DeAndre. "Have you been wondering where I go all day?"

"No sir—I already know," Wozniak replied. "I asked Anh one day and he told me. Personally, I think it's great that you're helping the Jarai. They've been screwed over so badly by Thiếu tá Hoang, who, by the way, hasn't indicated he's wondering where you are."

The subject then changed to their impending reassignments. Wozniak was anxious to see if he would get an "early DEROS", meaning go home ahead of schedule.

That evening, as DeAndre sat in his hootch, he wrote a long overdue letter to his mother, telling her of his impending transfer to Pleiku—and out of combat. "Mother, not much happens in Pleiku. It's a headquarters place, not a combat outpost. I think I'll only be there a short while before I'm transferred to another headquarters place. Mother—in other words—I'm safe."

But he felt his guilt. He was ashamed for not telling

her he was strongly considering both a relationship with a Jarai woman and joining FULRO. If his mother knew either of these ideas, she would freak out in anxiety. He didn't want his mother any more scared than she was already. As to Chel, his mother had always been open to him about falling in love and settling down, but he doubted she would be happy with his settling down with a tribal woman.

If she had reservations about him and Susan, how much more nervous would she be about Chel. As for staying in Vietnam, he knew his mother's reaction to that idea.

But before deciding about Chel and staying in Vietnam as a part of FULRO, he needed some more information. The following day, he was able to get Lee Anne, Smithy, and Chel together in the church building.

"I'd really like to talk to all three of you, and I would like it to be totally private. I wouldn't want anybody else to know we met, whether Jarai, Kinh, or American."

That raised Smithy's eyebrows. He also had to explain to Chel what "totally private" meant. Lee Anne seemed to have some idea what DeAndre wanted to talk about and just smiled. Once Smithy finished his explanation to Chel, she looked directly at the big black man, her face a mixture of pleasure and perplexity. She got up, walked across the room, and sat down next to him. Very close to him. Then she shyly reached for his hand and held it in hers.

Only Smithy seemed to be in the dark about the subject as he said, "I suggest we meet at Dr. Pat's hospital. If any of Thiếu tá Hoang's rangers see all of us there, that would not seem out of place. Kinh never go inside because it's for Jarai. When do you want to do this, DeAndre?"

"As soon as possible, and well before I go to Pleiku."

"How about tomorrow at nine in the morning?" asked Smithy.

"I'd prefer ten. That way the guards don't wonder why I am leaving the Ranger compound early," replied DeAndre.

They all agreed on the next day at ten. As the ladies were leaving, Smithy silently grabbed DeAndre's arm, steering him outside.

"What's all this about? And why the secrecy?" It was obvious Smithy was worried and concerned for his friend.

"I'm toying with an idea, Smithy. Just thinking—no decision made yet—that I might stay here in the mountains of Vietnam after the Americans leave. Yeah, I know I would be a deserter, but I'd like some ideas on how I could avoid that label. I don't want my mother to think I am a turncoat."

The light came on in Smithy's face. "Is this about Chel?"

"Not entirely, though she is a part of it. There is a lot more to all this than her. Smithy, I am just so damned tired of watching Hoang killing Jarai for no other reason than he is a screaming racist. I thought I knew a little about discrimination, but it's nothing compared to how I've seen the Kinh treat the 'Yards. But tomorrow, I want to talk to all of you about that being a reason to stay here.

Are my feelings legitimate? Are they enough to make such a drastic decision?"

"I'm glad you want other people's input. May I suggest one more person? How about Anh? He's not Jarai, but he would have some idea of the sacrifice you'd be making."

"Great idea! When I leave here in a few minutes, I'll track him down. Thanks for the suggestion. But there's one thing I want to ask you before tomorrow. Just you, because I would just as soon Chel does not know I asked."

"Okay. Fire away."

"When I was at the worship service with Chel, I told her I was surprised she was there because she is not a Christian. Her reply was that she is almost a Christian, but that to do that, she would have to give up the spirits. What does that mean?"

Smithy took a deep breath and said, "All the non-believing hill people trust that everything has a spirit; some spirits are good, some bad, some powerful, some almost benign. As an example, if someone gets sick, it's the fault of a spirit. If they are really sick, it's a bad and powerful spirit. The spirits are also jealous. If you don't pay attention to them, they get angry and cause bad things to happen. There are shamans whose job it is to figure out what spirit has been offended and how to appease it. In some cases, how to thank them. There are all kinds of rituals to follow. Some are really bloody."

DeAndre interjected, "Bloody? How so?"

"Okay. Let's assume a person in the village got sick but recovers. The spirits need to be thanked properly, especially if it was an important person. The family may offer up a chicken for a sacrifice, but the ceremony will involve the whole village. There is a lot of ritual involved and that ritual involves alcohol.

"In any house, you will see large jars. Those are filled with a kind of rice beer that is fermented in the same jar. Once the wort has begun fermenting, the lid of the jar is sealed. It is opened only for these rituals. The only time the Jarai drink alcohol is for a spirit ritual.

"When a ceremony begins, the jar is opened and water is added. Everybody drinks from the jars using straws. The poor chicken's legs are tied together while the people drink, and the local shaman says appropriate prayers.

"Then, the shaman picks up the chicken and sticks a knife in its beak so that it bleeds. The bird may or may not die. Very carefully, the bird is held over the feet of the healed person's family and a drop of blood falls on the feet. Eventually, the chicken's throat is cut and the blood poured into a bowl where it is mixed with some beer wort, making a paste. A bit of this is smeared on the forehead of the healed person.

"Of course, the chicken is not wasted. Once the person's feet and forehead have been anointed, the woman of the house will take it aside and roast it. The animal to be sacrificed varies according to the importance of the person. In some cases, a water buffalo is sacrificed."

DeAndre just gaped. This was all new. "Do the Jarai Christians do all this too?" he asked.

"No. That's what Chel meant. She would have to give all that up. It's not just leaving a religion. It is really the underpinnings of the culture. A person spends all his or her time thinking about the spirits. It's one reason why villages are usually either all Christian or all traditional. Strangely, there's no animosity between the two groups, though the shamans don't like the fact missionaries are putting them out of a job.

"A while back, just after Lee Anne and I met her, Chel's uncle came down with pneumonia. The local shaman could not cure him. Eventually, he was taken to Dr. Pat's hospital where he was cured. When he came back to the village, his family sacrificed a buffalo in thanks to the spirits. Chel had been exposed to Christianity, so the next time she was in Kon Tum, she asked me about how Christians would celebrate such a recovery.

"I told her we would have a meeting full of songs and prayer, thanking God for the recovery, but Christians do not kill an animal. That seemed to have an effect on her."

DeAndre was so flummoxed he merely sat there without responding. Smithy said nothing as he could see there was nothing the matter with his friend. He merely needed some time to process what he had heard.

Smithy waited a bit, then said, "Just to change the topic a little, you may have an 'in' with the Jarai. They believe that some people are favorites of the spirits, and those people are strong leaders. I think it very likely you will be considered one of those people."

DeAndre couldn't help himself and smiled. "I'm not so sure of that, and I'm not sure how I would demonstrate any 'leadership,'" said DeAndre, saying the word "leadership" derisively.

"Really, Infantry Captain Washington, who has two

combat tours behind him, doesn't know how he could be a leader among the Jarai? You don't think you could contribute to their efforts to fight the Kinh?"

DeAndre hadn't thought of that. It would be a very different kind of war than he had fought before, but he did know the basics of combat. But, right then, he also wanted to take his leave and get back to his hootch to chew on things.

With many hugs and thank yous, he left Smithy and the others, saying he would see them all in the morning.

He hadn't been back in the hootch for more than an hour when Anh came in. Feigning shock, DeAndre said "Well, this is a surprise. What's up?"

Picking up on the clue the captain gave with his head gestures that he didn't want Wozniak to hear what he was about to say, he replied, "Nothing, really. I had to go over to headquarters and thought I'd drop by. Wanna get a beer?"

With that, the two headed over to the tent that served as a club, got a Bier LaRue each, and went back outside. DeAndre quickly realized that Anh was up to date on all the latest ideas. "But I have one big problem. How do I join FULRO but not to be a deserter?"

Most likely Smithy had told him about the situation because it was obvious that Anh had already thought about a solution. As Anh laid out his idea, DeAndre's face broke into an ever-widening grin. After a few questions to fill out the details, Anh said, "Hell, Captain—we'll make you a hero—and not a dead one either."

DeAndre's face creased with unconcealed glee. "Wow!

You are a screaming genius, Anh!! That's perfect. But what excuse will you give your father for being in Pleiku?"

"Oh, no problem with that. The entire family is moving to Pleiku anyway. It's a safety measure. My father saw how weak the government was when Doan Ket was overwhelmed by the Communists in April. Pleiku is bigger than Kon Tum and has an American presence, even though the US has turned over control of the base to the ARVN. It's also on Route 19, making it easier to get to the coast if we have to leave the highlands.

Anh continued, "But seriously, are you really thinking about staying here? In my case, I could come home anytime I wanted, but you won't have that luxury. I think living in a Jarai village would be hard, as would leading them in combat."

DeAndre heard the comments, knowing Anh was playing devil's advocate.

"I appreciate the thoughts, Anh. I really do. Anything you can think of that might help me make the decision is welcome. Really."

"Okay. From what Smithy tells me, this has to do with Hoang being a war criminal and about Chel. Am I right about that?"

"Smithy didn't leave anything out," laughed DeAndre. "What do you think about Chel and me?"

"OOooo, that's a loaded question, DeAndre. I fell in love with a black girl at UCLA, but I admit that I chickened out when I thought about bringing her to Vietnam. Yeah, we could have stayed in the U.S., but I'd have to send photos and all that back home. I wimped out, so I'm a bad one to ask about affairs of the heart that cross racial and cultural lines."

DeAndre fell silent. His mother knew about Susan. They had never talked directly about Susan, but he knew his mother liked the woman but certainly wasn't keen on them marrying. What would it be like to bring home a woman who was from a very different race and culture?

Anh saw he'd hit a sore point. "Mind if I ask how strong your feelings are for Chel?"

"Good question, and the honest answer is that at this point, I really don't know. I feel very attracted to her. Besides being beautiful, she is also very smart. As Lee Anne says, we are waiting to see if something will blossom."

"Wow. That makes any decision even harder to make right now. So, going back to the original subject, is it worth it to stay here because of Hoang?"

"I'm chewing on that, too. I truly hate the man for what he is doing to the 'Yards, but staying here and living in a Jarai village for the rest of my life? It will get even stickier when the Communists win the war."

"You know, of course, that Hoang is merely a pawn of General Ngô Du, commander of II Corps. He is far worse than Hoang—far worse because he has almost total power here. He is the South Vietnamese government. Did you know that in 1966, when the US 4th Infantry Division was moving into the area, Gen Vinh Loc, the II Corps commander at that time, proposed displacing all the Jarai in and around Pleiku, about ten thousand people, from their villages to a new area further west, then have that declared a 'free fire' zone?

"He instructed his subordinates to look for FULRO members when they go into villages, and not worry about Communists. Ngô is not much different, though he is also into the drug business and smuggling. When Hoang does things like strafing Jarai with helicopters, he is just imitating his boss. More likely, he is trying to cozy up to Ngô."

"I can just imagine how happy Hoang would be if he captured me sometime," pondered DeAndre as he wiped dew drops off his beer can.

Finding his beer gone, Anh stood up, saying he needed to get back home and help with the move to Pleiku. "Let me see if I can sum it up. You're not sure if you're in love with Chel enough to never live in America again. You would have to learn to be a native Jarai. And there is a big chance of being killed or wounded fighting alongside the Jarai. Your best hope is that the highlanders win the war against the Vietnamese, whether north or south, establish a country, and get recognized by the U.S."

DeAndre had to chuckle. "Yep, that's a pretty good summation, Anh. Let me know before you move to Pleiku so I can catch you up on all this."

With that, the two men parted ways... with DeAndre no closer to making a decision.

In the morning, as he and SFC Wozniak walked to breakfast, they discussed their immediate future. "Well sir, my orders came through last night. I leave for Pleiku the day after tomorrow, then on the Cam Ranh to board the freedom bird and go back to the world."

"Well damn, Sergeant! Good news indeed!" as DeAndre clapped him on the back. "Sounds like a call for a few beers tomorrow evening, eh?"

The conversation all during breakfast was about their plans. "By the way, Thiếu tá Hoang knows we're leaving, but does he know when?"

"Ooo, forgot all about that. Thanks. I'll go see him today and let him know."

With that, DeAndre reached into his pocket and pulled out some piasters, handed the bills to Wozniak saying, "I buy, you fly." Smiling, the senior sergeant said, "And I know just the place to get the beer."

DeAnde walked to headquarters asking to see Thiếu tá Hoang. After being ushered in, he explained he would be leaving soon, though wasn't sure exactly when. "You good advisor, Đại úy. You teach me good things," said Hoang as he stood to shake hands with DeAndre.

"And I enjoyed working with you," lied DeAndre. "Perhaps I will see you again after I go to Pleiku." He fervently hoped he would never see Hoang again.

Knowing Wozniak was off somewhere buying beer, and Hoang not planning any operations, DeAndre spent the rest of the day in town with the Jarai. He became more used to Chel's displays of affection and realized he was falling in love with her.

"Yeah, we have eyes, DeAndre," said Smithy with a toothy smile as the captain left to return to the camp.

That night was a time of camaraderie as the two soldiers swapped stories and enjoyed both the beer and each other's company. If there is good chemistry, senior level sergeants and company grade officers often build strong relationships. That was the case for these two professionals, and for that reason, it was hard for DeAndre to keep quiet about his possible plans with the Jarai, FULRO, and Chel.

The beer set him up to reveal some of his thoughts. "Sergeant, have you ever heard of FULRO?"

"Yeah. It's some sort of Montagnard organization that wants to fight the Vietnamese. I think a lot of them were trained by our Special Forces."

"Bingo. That's it. They want the mountains to be an independent country."

Wozniak thought about that a little, then said something unexpected. "You don't suppose some of the Green Beret types are still working with them, do you? I mean, against the Vietnamese."

That caught DeAndre off guard. "I suppose there could be. I can certainly see why they would want to stay here with the Yards."

Wozniak pondered that answer for a moment, before turning the conversation towards talking about stateside assignments.

The following morning, the two walked to the American helipad as a Ranger brought the NCO's meager belongings. During the short wait for the bird, the two promised to stay in touch. When the helicopter landed, Sergeant First Class Wozniak left his duffel bag on the tarmac for a moment, faced Captain DeAndre Washington, came to a position of attention and rendered a hand salute that would have delighted any drill sergeant. The captain returned in kind, both showing their respect for each other. After the two dropped their hands, Wozniak picked up his duffel, ducked under the whirling blades and clambered aboard. As the bird lifted off, the two exchanged waving hands and smiles.

CHAPTER ELEVEN

He could now spend all his time in Kon Tum, which he did, but he was still careful that no Rangers saw him with the Jarai. His efforts shifted from rebuilding the shattered Jarai enclave to helping Smithy, Lee Anne, and Chel prepare for moving to Pleiku.

As typical westerners, the American couple had a lot of things to arrange and gather. They also had to say goodbye to the Christian Jarai they loved deeply. Many tears were shed at the last worship service, and one of the members of the congregation led them all in prayer for Smithy and Lee Anne.

As for Chel, her life of simplicity brought on by the semi-nomadic ways of the mountain tribes made her preparation for moving very easy. She was ready in fifteen minutes. But she did go into the jungle to see her parents in their village for a day to tell them she was going to Pleiku. She wasn't as downcast as DeAndre expected her to be. "Pleiku not far. I see them and village again."

A civilian truck had been hired to transport household goods, as well as some church objects. Once it had been loaded, DeAndre brought the Americans onto the base to sleep the night in his hootch. Up early in the morning, and after breakfast at the Vietnamese mess hall, they went back into the city. As last-minute goodbyes were being made, DeAndre worked up his courage and signaled Chel to join him at the back of the church building.

"Do you know the word kiss, Chel?"

"Yes. Please."

The long lingering kiss extended to many more kisses and included a few groans from both of them. The honest passion was evident in their eyes. They broke apart only when hearing Smithy call Chel, saying it was time to leave.

As they emerged from behind the building, Smithy spotted them. This time there was no teasing grin. Instead, he gave them a smile and an approving nod of the head.

"DeAndre, got a minute to talk?"

"Sure."

Chel knew the two men wanted to be alone, so she wandered off to find Lee Anne. "I think you have the issue settled as far as Chel is concerned," said Smithy. "But that's only half the equation. Before you make a final decision, I strongly suggest you find out more of what it might be like to live as a Jarai. Chel's home village is close. I suggest you go there for a day and see what it's like. It would also give the people a chance to meet you and see how they would handle a black American in their midst. Living in a Jarai village is going to be very different from anything you've experienced."

"Is that a warning, Smithy?"

"Very much so."

The two men went off in search of the women. On

finding them, DeAndre took Chel aside. "Smithy says I need to see what living in a Jarai village will be like. Can we visit your village before you go to Pleiku?"

"Yes. I like. You think maybe you stay one night?"

DeAndre could tell she wasn't suggesting anything physical between them but rather saying it might be a good idea to spend a 24 hour cycle there.

The idea of being discovered by the Rangers was a bit unsettling. If Hoang had any inkling that DeAndre was with the Jarai a lot, there was a very good chance he would be assassinated by some Rangers before going to Pleiku. He hadn't noticed anybody watching him, but then again, how would he know? He did think the likelihood of him being watched at night was slim and all Hoang had to do was ask the guards. But the guards did not keep a log book, and the shift changed during the day. If he didn't come back on base in the evening, they probably wouldn't realize it.

He made a snap decision—I will go with Chel the day after tomorrow. They would leave in the morning, it would take about two hours to get to the village, stay the night and return the following morning. He'd tell Hoang he was going to Pleiku and would spend the night.

Chel beamed when he told her the plans. Smiling, she said, "Wear real clothes, not Army clothes. Today, I ask boy in Kon Tum to go village and tell them we coming and tell them you want learn Jarai." She was so happy she took his arm and jumped up and down with glee.

After a brief hug, she went off to find the others. Luckily, Anh was with Smithy and Lee Anne. He told them the plans. Lee Anne looked worried, Smithy seemed happy, and Anh thought DeAndre was nuts. The day was spent working on both the church building and the homes of some of the Christians who had lost everything during the fight over Kon Tum.

"Lee Anne—what will your church members say if they see me give Chel a quick kiss?"

"Good question. I don't really know, but I suggest you just give her a kiss on the cheek. That might go over better."

As the afternoon wore on, he worked and thought, wondering if he was doing the right thing. He realized that as long as he wasn't discovered, he would still have time to change his mind once he returned from the village. The risk was real, but slight, so he decided to go through with the village visit on the next day.

Chel and DeAndre set out an hour after sunrise. After fording a small river by walking on some well-placed rocks, they were quickly in the forest. DeAndre could barely discern the trail, but Chel followed it as if it was a four-lane highway. As per Chel's suggestion, he wore blue jeans, a dark long-sleeve shirt, and sneakers. The only military item he wore was his bush hat, but those were not uncommon among civilians. He felt naked with no weapon—he was most uncomfortable with that idea—but Chel told him that if he carried a gun, the Jarai would wonder if he really was an anthropologist.

Chel was dressed in a traditional Jarai skirt and tunic. He also realized that though he had been in the forest many times before, it had always been with troops. It was much different now when there was just the two of them. He could hear all the sounds of the forest, feel the breeze, and enjoy the walk without wearing a pack.

A tiny little hut appeared, looking abandoned and bedraggled. Chel explained that a leper who was in the terminal and most contagious stage of the disease, had been banished there as a way of protecting the others. Chel went on to say that the banned were not being persecuted as the family would bring him food and provide company for a few hours each day.

In some rare cases, a victim's spouse might live with the leper. After the person's death, the hut was left standing, primarily because nobody wanted to contract the dread disease by tearing it down.

Approaching the village, Chel asked him to wait under a tree while she went ahead to see what was going on. Returning quickly, she said, "Very big thing. They number two day of buffalo killing. Many things happen yesterday. Today they kill."

They walked together into the village, greeted by a combination of stares and smiles. One of the men arose and said something to him. DeAndre answered in his halting Jarai, thanking the chieftain for the welcome. This unleashed a torrent of shouted surprise that this black man could speak their language. Seeing he was overwhelmed, Chel interjected that he was a beginner with Jarai. With that, DeAndre sat at a respectful distance, took out his notepad, and took some notes.

Chel joined her clan, which included her parents. With the proper decorum expected of a daughter, she brought them to DeAndre, making introductions. Her father was dressed only in a loincloth, and her mother was bare breasted.

At this point, DeAndre and Chel had agreed they should only know him as an American who wanted to learn more about the Jarai. He stood as they approached and kept standing as she told him who they were. She had coached him, so he recognized many of the words. Then she introduced him. At Chel's cue, he spoke in Jarai and thanked them for allowing him into their village and that they were most gracious. Surprised he spoke some Jarai, they beamed their acceptance.

After they walked away, DeAndre had no clue what was going on around the village, but he figured he should look like he was trying to find out since he was supposed to be learning about the Dega. He figured he could ask Chel later. Resuming his note taking, he sat back down.

In front of him was a longhouse, made of wattle walls and a thatched roof, and easily the biggest house in the village. Since it was the most imposing, he assumed the chieftain lived there. Across the informal courtyard formed by the houses stood another building, quite a bit taller than the others with a very steep sloped roof. DeAndre wondered why all the houses, including the longhouse, were built on stilts. The little stream nearby looked too innocent to cause a problem.

Set aside from the rest of the village was a small building, too low to be a house, and surrounded by a wooden fence. Perched on the fence were carved wooden statues which included couples embracing, pregnant women, and people in mourning, along with elephants, and birds. There was even one man holding his erect penis in his hand. Chel told him later that the building was a burial vault.

He also noticed there were no chairs in the village. Everyone either sat crossed legged or squatted. DeAndre realized this would be a problem if he stayed with the Jarai. He couldn't squat like they did, sitting flat footed on their heels and he knew his long legs would quickly go to sleep if he sat with them coiled underneath him for very long.

There was a long platform in front of the house but built directly on the ground. It was about the same length as the long porch on the house and was crowded with both men and women. There were a number of large clay jars, some still sealed but others open and surrounded by people using bamboo straws to drink.

DeAndre expected to see all of them clad in native clothes, so he was surprised to see some of the men in western clothes, some in loin cloths, and some a combination of the two. It had a strange effect seeing the two styles.

He noticed an older man approach Chel and point towards DeAndre. She smiled, then came back to tell him it was indeed the head man's house and much of the ritual involving the sacrifice would be at the long house. Normally, chickens, dogs, pigs, and other animals are sacrificed. A buffalo is very expensive and only the rich and powerful could afford one. The leader of this village is uniting in a political alliance with another village chief and the sacrifice today was necessary to have the spirits bless the union. After a brief introduction and a thank you from DeAndre, the man left to attend to his duties.

"What's in the big jars, Chel?"

"Maybe like rice beer. Women crush rice. Make thick juice. Put in jars and covered. Juice is beer in maybe week—maybe two week. Jars not cover off until ceremony to spirits, then uncover and water in. Only time people drink, how you say, the thing in beer that make you happy."

"Maybe you mean alcohol, Chel."

"Yes, yes. Also, much praying to spirits with beer. Spirit worship only time Jarai drink alcohol."

After a concealed smile, she returned to the village group, taking a few turns at the straw as well.

DeAndre noticed a long bamboo pole, taller than most bamboo, and very straight. It was laying on the ground near the end of the porch of the house. All the leaves and branches had been trimmed off. When the man who had welcomed him to the village came out of the house, accompanied by a younger man, the crowd quieted and watched carefully.

The chief sat down in front of a jar and his companion brought a bowl of wort and a young chicken with its feet tied together, setting these between the chief and the jar. As the older man began to drink from the jar, the helper began to talk in such a way that DeAndre knew he was praying.

Basically, the prayer was to offer the rice beer to the spirits. Soon, another older man appeared in the doorway, the man whom DeAndre assumed was the other leader, sat at the jar and drank. Fifteen minutes or so went by before the local chief picked up the chicken and slit its throat, draining the blood into the bowl of wort. More prayers were raised.

Soon, the four arose and went down to the long-rimmed bamboo pole. Taking a bit of the blood and wort mix, each anointed the pole, all the while offering prayers. Next, they approached a big bowl of white rice and anointed that as well. Finally, the chief walked back to the entrance to his home, and once again, smeared bloody wort on the doorway, while uttering a very long prayer.

Finally, two large jars of beer were opened. Obviously, there was some sort of rank or protocol to be followed as there wasn't a rush to the jars. Each person, both men and women, drank four gulps from the jar before ceding their place to the next person. DeAndre saw the chief approach Chel again and point towards him.

"Man wants you drink rice beer." With that Chel escorted DeAndre to one of the two jars. When it was their turn, she said, "Four drinks." The diluted rice beer was nowhere near as potent as he expected it to be, but his biggest feeling was one of gratitude for being asked to join in the activities. He finished and looking at the chief, smiled with a slight dip of his head. The smile was returned.

The drinking continued and the conversations got louder. DeAndre began to understand that the amount of beer consumed made up for its lack of potency. Finally, in the late afternoon, a second chicken was brought forth. While still alive, the chief used a knife to crack its beak. The blood was dripped onto the feet of the newly arrived people from the other village. As before, the chicken does not go to waste as some of the women defeather and dress the birds for later consumption.

Night fell and the drinking continued. Chel told him that it would not be proper for her to sleep near him, but she would stretch out to the left of the lower platform if he needed her. DeAndre had slept on the ground many times before, so he found a rock less and rootless place, used his small backpack for a pillow, then pulled his poncho liner over his body. He let sleep wash over him despite the noisy carousing at the platform.

Dawn was announced by the ringing of gongs on the big house's platform. Even though the sun was still below the horizon, the chief called for the revelry to end. Soon, there was enough light for some young men to be seen leading a large water buffalo out of the forest. Moving with the slow lumbering pace characteristic of his kind, the animal finally came close enough for DeAndre to see a kind of floral crown on its head.

Near the house, the long bamboo pole that had been trimmed and anointed yesterday was now standing in the ground. It was short enough that DeAndre knew it had been very deeply planted. It had also been decorated with what he assumed had something to do with the spirits. Knowing the Jarai had no written language, the symbols seemed to be artfully rendered graphics.

The young men used a rope tied to a ring in the buffalo's nose to guide him to the pole. After tying the animal to the stake, they secured another rope to the animal's nose ring and tied that one to a different place on the pole. The gongs continued which mildly agitated the buffalo.

Suddenly, the noise stopped and the chief's wife appeared on the platform. She took a handful of glutinous rice from a basket then brought it to other women on the platform. Each in turn, took a handful, ate it slowly and thanked the lady of the house. When all had been served, she raised the basket in the air while the chief said a long prayer.

With that, the gongs resumed tolling, but now, much louder, agitating the buffalo even more. After presenting himself to the chief, a young man armed with a sword then bowed in front of the animal, saying a prayer. Walking behind the unnerved creature, the swordsman suddenly leapt forward. With strong strokes, he severed the right rear leg tendon, then in another lightning move did the same to the left tendon. Maddened with pain, the beast collapsed on his haunches while trying to arise by thrashing his legs, smattering the nearby crowd with blood. Within seconds, a second young man delivered a fatal blow to the base of the animal's skull. Even before the buffalo had fully succumbed, the swordsman quickly slit its throat while the other placed a metal pan under the slit, capturing the gushing blood.

Someone cut off the animal's tail and took it to the village shaman. Someone else cut off his testicals, giving them to nearby children as a treat.

The festivities now turned away from the dead buffalo and resumed with the chief saying another prayer and inviting the crowd to drink. Young men and older boys were seen dragging the body away to wash it in the nearby creek. The village women would soon butcher the carcass, giving the prize cuts to the important people, but also ensuring everyone in the village got some.

Chel returned to see a noticeably stricken DeAndre. She hadn't expected this. She knew DeAndre had fought in many battles and therefore would not be bothered by this at all. She had been raised attending these sacrifices and they were completely normal to her.

"You no look good," she said gently.

He said nothing at the moment, then finally asked her to take him back to Kon Tum.

She was very puzzled by this, but merely said, "Okay."

= = = = = =

"Smithy, I don't think I could live in a Jarai village the rest of my life. That was the bloodiest damned thing I've ever seen." He had asked to talk to Smithy as soon as they returned to the church building but also asked Chel not to leave the area while he talked to Smithy.

"It is pretty gruesome, DeAndre. I get that. Frankly, it's less about religion and more about culture. It's all to keep the spirits happy."

"You told me the story about how Chel's uncle was cured at Dr. Smith's hospital and that had a strong influence on her. Why don't such healings affect all the Jarai that way so that they stop their barbarity?"

Smithy took a while before answering. "That was because of something that happened to Chel and her family. Few people in this world make big decisions based on what happened to anyone else. But don't forget—you have alternatives to living in that village."

"Like what?"

"Like living in a Christian village. There are some believer hamlets in the jungle around here."

"Will Chel take me?"

"Sure, but you need to ask her now so she can notify the village."

"I will. One more thing, Smithy. Chel has no idea why I was so grossed out. My Jarai is not good enough to explain it to her, and neither is her English. Could you explain?"

"Sure, but let's do it right now so I can answer any questions she might have that I can't answer."

DeAndre brightened up. Smithy's idea was spot on. With a big smile on his face, he went to the door, and by wagging his fingers in the way that most westerners use to wave goodbye, he asked her inside.

Smithy first asked Chel if she had any idea why DeAndre wanted to leave the village so quickly after the sacrifice. Replying in Jarai, she said she really didn't know why he'd reacted so negatively. She thought that because he had seen so much combat, the blood wouldn't bother him. Smithy relayed that information back to DeAndre, then asked why his fighting experience hadn't prepared him to see the gore.

"Smithy, people don't always know what happens in combat. Yes, I've seen a fair number of dead soldiers, both American and Vietnamese, but I never saw anyone die by my own hand. I didn't care if they were dead or not. I just wanted the enemy neutralized so he couldn't hurt me. I most certainly never tied up a Communist and ceremoniously killed him with the intent of catching his blood in a pan!"

He realized after saying the last sentence that he had fairly shouted it. "Besides, Smithy—much of the time people are killed by other people who have done something that almost asks for getting killed themselves. You know, like charging a bunker yourself or running out in an open place where the enemy has an easy shot at you. That poor buffalo never did anything to ask for that. Yeah, I know the meat didn't go to waste, but cattle slaughtered for meat don't have their rear leg tendons sliced with a sword either."

He realized Chel was watching him closely and knew she probably understood a fair amount of what he said. Nonetheless, Smithy took his time to translate all that into Jarai.

Smithy looked at DeAndre and asked, "You want me to ask her about going to a Christian village?

"Please."

Chel's face gave away her answer. DeAndre's Jarai was good enough to understand that she would ask some of the church goers, some of whom had lived in a village rather close to Kon Tum. Throwing a glowing smile towards DeAndre, she left quickly.

Within an hour, she returned saying she had talked to some of the village dwellers, all of whom said they would be very delighted and honored to have the American visit.

One man came forward and declared his business in Kon Tum was finished and he would return to his village early in the morning.

Lee Anne arose well before first light and fixed DeAndre breakfast, who had gotten up even earlier, so he had time to walk from his hootch, past the sleeping gate guards to the Smith home. The first rosy streak of light filtered through the forest trees just as Chel arrived. Before DeAndre hoisted his backpack, Smithy asked both of them if he could pray for them. Both readily agreed.

Placing his hands on their shoulders, Smithy intoned, "Father, we all know how important this visit to a Christian village is for both DeAndre and Chel. We ask that you keep them safe, returning here to decide their futures with your help. In Jesus' name, Amen."

With that, and hugs from Lee Anne, they started off. This village was closer than the other and they arrived by mid-morning. Their guide asked them to wait before entering the village so he could talk to the people. Returning a half hour later, he said they both were invited in.

As they approached, DeAndre noticed the village was smaller than the previous one, with only two buildings: a large longhouse and a much smaller but taller building. Both were built with the same wattle walls and thatched roofs, and both built up on tall sturdy stilts. In many ways, the two villages were the same: the dress was the same, the houses the same, the cooking fires the same.

But there were some differences too. For one, there was no grave house surrounded by carved wooden statues. There were no bare-breasted women and fewer men wearing loincloths. Many men were wearing western clothes, though they certainly knew nothing about fashion. One came out of the longhouse wearing American military fatigue pants, but with a bright red Hawaiian shirt and a black beret.

Toddlers still didn't wear bottoms. Walking alongside the longhouse, he looked through an open door and saw an AK-47 hanging on a peg. DeAndre quickly realized this was a FULRO village as well as a Christian one. Conspicuously missing were the large clay jars of alcohol used during the ritual sacrifices. But he did notice a few smaller jars which may have held some rice beer.

He was under scrutiny and he knew it. Though people smiled at him, they also had a questioning look on their faces. DeAndre had his suspicions confirmed when Chel told him they had never seen a black person.

Except for one man who stepped out of the tall house. Wearing khaki pants with a wide belt and huge belt buckle, plus a t-shirt with "Florida Southern College" emblazoned on the front, he confidently strode towards DeAndre.

"Welcome to village," he said in halting English. "I Siu Der."

And with that, DeAndre came to know the village chief. In his weak Jarai, he thanked Siu Der and gave his own first name.

This pleased the chief as he burst into a huge smile saying, "You speak Jarai. Only white man who know Jarai be soldiers." It was then that the man noticed he had used the phrase "white man" with DeAndre and he quickly stuttered, trying to say he was sorry. Chel stepped into the conversation, telling Siu Der that it was okay. DeAndre sensed what she was saying and nodded his affirmation. "Yes, yes, mean American soldiers."

Changing the subject quickly, Siu Der said that the villagers were about to have a morning prayer session inside the tall building and asked DeAndre to join them. "Of course," was his genuine smiling reply. He watched as many, but not all, the villagers entered the building. "Some people do other important things," said Chel, "but most come tomorrow or other day."

Siu Der excused himself, climbed the ladder and went inside. Chel and DeAndre followed soon after. They were met with a large cross suspended between the walls towards the front of the single large room. Off to one side was a small podium that might be used by a worship leader. The only decoration on the walls was a single small painting of Jesus sitting and talking with children. Straw mats were spread on the wooden floor and families took their places.

Chel smiled inwardly as she watched DeAndre slowly and tentatively get down into the cross-legged sitting position. She had been around Lee Anne and Smithy enough to know that westerners always sat on chairs. As she sat down next to him, she was aware of the eyes watching the two of them.

Soon, the singing started. If people had been wondering what kind of relationship the big American and the non-Christian Jarai girl had, they were immediately surprised to hear Chel's voice singing the old hymn

"Amazing Grace," in Jarai, of course. DeAndre sang too, but much lower in volume because he was using English lyrics. The people sang without accompaniment. Someone gave a starting note, and the rest took it from there. DeAndre delighted in this singing. It was joyful, harmonious and heartfelt. He couldn't help but think of the little white church back home.

He liked that members of the congregation led different parts of the service. One man stood to read from a book, which DeAndre assumed was the Bible. Someone else stood and read from another part of the Bible. By the call-and-response used, he knew he was listening to one of the Psalms. Next, another man rose, and standing behind the little podium, he lifted the Bible into the air towards God, and the group all got to their feet.

After the reading and everyone seated again, the leader spoke for no more than 3-4 minutes. Everyone's eyes closed, and a sense of reverence fell on the people in the tall wattle hut. People spoke in turn, alternately raising their hands and then putting their faces in their hands. It was obvious to DeAndre that this was the prayer portion of the meeting, with words of adoration and petition.

Within ten minutes, everyone who wanted to had made their prayer and the room was quiet. DeAndre peeked out of one eye and noticed all had their eyes closed. Finally, the leader—the one who had read the Bible from behind the podium—spoke softly. The people rose, and with their hands in the air, joined the leader in a benediction. Again, DeAndre felt a little like he was in church back home, except people spoke a different language.

Once outside and back on the ground, people approached Chel, presumably to ask her who the big black man was and why they were in the village. DeAndre stood silent, but when he heard someone ask Chel if he was a Christian, DeAndre surprised them all by answering himself, telling them he was. Startled at his language skills, they beamed their approval.

Slowly, the people left the two alone and went about their daily business. At first, DeAndre just sat with his back to a tree, watching and absorbing. There were a few fruit trees, chiefly mango and banana, scattered about. Off away from the long house were a few pig sties. Getting up, they strolled to the other side of the longhouse and saw small chicken coops along the skirt of the building.

Entering the long house, he was in a common room that was about one third of the length of the entire house, along with smaller rooms belonging to families. It was very dark, but when his eyes grew accustomed to the gloom, he saw that there was one room bigger than the others. DeAndre assumed it belonged to the chief and his wife. Near the door to each room was a granary for rice.

But DeAndre realized another big difference between the two villages when he spotted three men getting ready to butcher a pig. While two of them held the pig, the third man lifted his arms towards heaven in obvious prayer. Chel told him they were thanking God for the food. Then one of the men quickly slit the pig's throat, killing it in the quickest, most humane way they could.

DeAndre wondered if the village had ever been raided by ARVN troops, including the Rangers. He and Chel found Siu Der working in the field, tending the crop of upland dry rice. They had decided Chel would ask the questions so that the language barrier would not cause a problem.

"No, they never raided because we not near road. Southern soldiers not go far from trucks. We too deep in jungle."

But, since the village was made up of people who had lived in other villages before becoming Christians, Chel asked if anybody knew of villages that had been attacked. "They be Christian but hate Kinh. Many family killed by Kinh soldiers during raids. That's why they in FULRO."

DeAndre asked if she knew what a pacifist was. She did not. "It's a person who does not believe in fighting wars. They refuse to kill other people."

"People not pacifists. Want the Kinh out of the Highlands. They want dega nation," she answered. DeAndre tucked it into his mind to ask Smithy about the idea of a Montagnard nation since there were so many tribes and languages.

Siu Der broke away from his work to talk to the pair, wanting to know whether they wanted to spend the night in the village. Chel and DeAndre talked briefly before deciding they had both seen enough to go back before dark. After they told Siu Der their decision, he smiled and said, "Hope see you more. Village like you both."

The uneventful walk back to Kon Tum was quiet, neither wanting to make any noise that might be overheard and their safety compromised. The sun had set but there was still enough light to see colors when they passed by the first building in Kon Tum.

Lee Anne wanted to prepare a celebratory feast for them all, but DeAndre reminded them he had not yet decided. Besides, he wanted to ask Smithy about Chel's use of the term, "Montagnard nation."

"Oh yeah. She's spot on. As I said a while back, the fancy academic word is ethnonationalism. If you notice, the Montagnards do not fight amongst themselves. They intermarry regularly. Food, housing, manner of dress, farming methods are all common among the tribes. At the tribal leader level, they have even adopted a common language and speak Rhade—a big tribe south from here."

"Is that what FULRO is about?" asked DeAndre

"Yep, they want a free and independent nation here in the mountains of Cambodia and Vietnam that is recognized by other nations. They want to be free of the Kinh."

"Do they really think they can do that?"

"Most FULRO fighters were trained and armed by American Special Forces. Because of that, they believe the US will be on their side."

With that, DeAndre returned to the Ranger camp and his own little hootch. He knew he had more time. All of them would be going to live in Pleiku very soon, and he really wouldn't have to decide until he was reassigned from Pleiku and had to leave.

Nonetheless, it was a restless night. Certainly, the FULRO village was much better than the traditional one, but he wondered if he could live there the rest of his life. That would be one hell of a big culture shock. He did realize he was coming closer to saying he was in love with Chel, but not quite sure they should wed. He also knew he had time to ask the Army to marry her and take her back to the US, but then, that would be asking her to endure a tough culture shock with never having been able to see beforehand what America is like.

But DeAndre still had two nights left in Kon Tum before leaving for Pleiku. On the next to last day, he worked among the Jarai and asked Anh to come to the hootch that night. DeAndre wanted to pick his brain about how he could stay in Vietnam. Furthermore, he would supply the beer. That was an offer Anh could not refuse.

"I think the thing I fear the most is that my mother will be told I am a deserter. I don't want to do that. She will know I could not really be a deserter, but then again, she'll have no idea what happened or where I am. You have any brilliant ideas, Anh?"

It was quickly apparent that Anh did have a brilliant idea. Lubricated by the beer, the two tied up some loose ends in the plan and realized it was pretty good. All that was needed was to get Chel in on it.

CHAPTER TWELVE

DeAndre's departure from Kon Tum a few days later was super simple. The hootch and everything inside it belonged to the Rangers, so there was nothing more for him to do but gather his own personal belongings, his gear, weapons, and uniforms plus the bottle of scotch, then pack it in the duffle bag. Going through the Ranger Sergeant Major, he had arranged for a jeep to pick him up and deliver him to the American helipad. LTC Arredondo had sent a jeep to pick him up. Within two hours of walking out of the Kon Tum hooch, he was in the senior advisor's Pleiku office learning the ropes. It would soon be his office.

"There really are only two people you need to meet. First is the province chief, Pham Vu. He's a pretty nice guy. The other is the commander of II Corps, Lieutenant General Ngô Du. If you thought Hoang was a son-of-a-bitch and crooked, he's a rookie compared to Ngô. But you gotta play ball with him, so just grin and bear it."

DeAndre told his boss all that Anh had told him, as well as telling him Smithy had seen Ngô on a village raid when civilians had been murdered. "Yeah, I'll be nice, but not sure I know why."

Arredondo didn't rebuke him, knowing Captain

Washington was a good officer who would get the job done.

The following day, the colonel took DeAndre to see both Vietnamese officials. Vu did seem like a nice man and he spoke excellent English. That would make communication easy.

Ngô was another matter. He clearly did not like having to interface with a mere captain. It had been beneath his dignity to deal with a lieutenant colonel, but a captain? He made his displeasure known to both Arredondo and DeAndre. Ngô knew the black man had been Hoang's advisor, having seen him at a briefing at the Ranger's headquarters.

The ARVN general decided he would intimidate the American captain by telling him "the truth."

"With me in helicopter now," said Ngô.

"Yes sir," replied DeAndre.

Within minutes a jeep arrived to pick them up, leaving Colonel Arredondo and his driver alone. On the way to the helipad, the jeep stopped to pick up another man. He was an ARVN sergeant who introduced himself as Quan, Ngô's interpreter. He explained. "I learned English at Fort Benning during the Noncommissioned Officers Candidate Course."

"Pleasure to meet you, Trung si," said DeAndre. Of course, he knew Fort Benning well.

As the jeep approached the helicopter, Ngô and Sergeant Quan chatted for a few minutes. "He wants to take you up in his helicopter to show you the area and give you some idea of what to expect in your new job."

"I don't know about your knowledge of recent Vietnamese history," Quan began, "but in 1954, when the French were expelled, the Communists controlled the north, while the south was free. There was a three-hundred-day period when people could travel freely. Almost all the Catholics in the north feared persecution, so they came south. There were well over three hundred thousand of them. The south welcomed these refugees of course, but there was still the question of where they would go and what they would do to become a real part of the society."

With that, the party climbed out of the jeep and boarded the helicopter where they all put on helmets that allowed them to talk on the intercom. As the bird lifted off, Quan resumed, "The hill people did not take care of the mountains. They cut down large parts of the forest and let the trees rot. They themselves were, and still are, a very uncivilized people who are also very poor. It quickly became government policy to assimilate these savages. One of the best ways to do that was to bring agriculture to the highlands. Since many of the northerners were farmers, it was decided to build entire villages up here and have them grow coffee and tea."

DeAndre recognized propaganda when he heard it but did his best to conceal his contempt from showing on his face.

"It is hoped the new villagers can teach the savages how to farm. But they resist us. The Americans not only trained them as guerillas but gave them lots of weapons too. It is one of eneral Ngô's duties to subjugate them as per official government policy. The general takes his orders directly from President Thieu."

The helicopter banked and DeAndre saw a village on the edge of the forest. Near it was a stand of very short and young coffee trees. "This is one of the new Catholic villages. It is our duty to protect it."

Then the bird gained some altitude before flying over the jungle. "The pilot does not want to be a target for the savages, so he has gone to a high altitude."

With that, they went on for a short time before the helicopter banked again. This time DeAndre caught a very brief glimpse of what looked like a longhouse. "This is one of the savages' villages. Do you see the large, cleared area to the north that looks like it has recently been burned? That is how they are ruining the forest and the mountains."

The bird soon began flying back to Pleiku. General Ngô turned to his new contact with the Americans and said, "Like French, we *civilisatrice* the savage." DeAndre found it amusing hat the Vietnamese were using the same rationalization as the French to conquer other people.

DeAndre was a bit surprised, but the helicopter began to lose altitude. "We're going to observe an army unit working," said Quan. "That's why the pilot is not afraid that the moi will shoot." Quan had code switched to using the Vietnamese word for savage.

The pilot skillfully landed the aircraft in a nearby landing zone, and the party was met by a small contingent of Rangers who escorted them to the village. Ngô stopped far enough away to watch, telling Quan he didn't want to interfere with the operation. Off to one side, DeAndre recognized Thiếu tá Hoang. He realized Ngô had come here specifically so the American could see Hoang in action.

DeAndre knew the general was watching and that he wanted to show off a little to the American. The perfect moment arrived when some of the Rangers found two young Montagnard men in a tunnel, both armed with M-16s. Bringing them to the battalion commander, the men were turned over to Hoang's bodyguard and the M-16s put in Hoang's jeep.

The Rangers ordered the two prisoners to kneel, then kept them in that position for a long time, not allowing them to sit on their heels as they normally would. DeAndre could see the pain on their faces.

An hour later, and not finding any more armed men, Hoang gave orders to end the operation and for his troops to return to their trucks. With that, he turned to the two prisoners, pulled out his sidearm and shot each one in the forehead at close range. DeAndre was very much aware that Ngô was watching closely. He thought he was successful in masking his feelings.

After an exchange between Ngô and Quan, DeAndre was told, "FULRO is more dangerous than the Communists. That's why General Ngô orders his subordinates to prioritize FULRO."

The return flight back to Pleiku was uneventful and only a few words were exchanged. The bird landed and DeAndre's lesson in brutality was over. He thanked the general for his hospitality and Sergeant Quan for the information, smiling as he did so. He thought he passed the test.

As they rode back to the office, Arredondo repeated his feelings about Du, and as before, knew DeAndre would handle it well.

The office was welcoming as it had air conditioning, as did DeAndre's new sleeping quarters. "But the small staff we have here is leaving with me except for a driver who will double as radio operator. You'll keep this building, but for everything else, you'll need to go to the aviation area. My guess is that you will leave before they do, though there aren't many of them here anymore."

With that, the two ambled over to the helicopter company's mess hall for some good old-fashioned American hamburgers with a side of baked beans. "Your driver, Spec Four Jason Cabot, is new, sent here when the last battalion advisor to the Marines went home. Not a bad kid and does his job well. If I were you, I'd ignore the smell of marijuana that emanates from his hootch at night."

That brought a chuckle from DeAndre. "Not to worry, sir. If the kid is doing his job, what the hell do I care if he's getting stoned. Let's face it. In Vietnam, all of us choose some form of self-medication. The officers have scotch, the NCOs have beer, and the lower enlisted personnel have weed. Sounds like a deal to me."

"Ah, you are an enlightened man, DeAndre."

Officially, the colonel and DeAndre had two days to transfer duties, but they found it didn't take anywhere nearly that long. He would have no staff, such as an intelligence officer, nor a liaison officer to the Province Chief that would allow him to coordinate artillery fire.

"I'm not joking, Captain. You really have nothing to do except stick your neck in Vu and Du's offices once in a while and say nice things to the advisory folks back at Cam Ranh."

After spending the morning in the office and getting to know Sp4 Cabot, they were free to pretty much do as they wished.

On the morning of Arredondo's departure, Cabot drove both to the airfield to catch the Caribou that would take the colonel to the big base at Cam Ranh Bay. After the customary exchange of salutes, DeAndre returned to his almost empty headquarters and found he had mail. It was from Sergeant Rocatta.

"Sir: I want to apologize to you. While I served under you, I was only courteous enough to meet regulations. I was raised in a very racist home in Ohio. My father used the N word as a matter of course. I carried that way of thinking into the Army. It makes me nervous to actually write this, but I hated working for a nigger captain.

"But, I don't use that word anymore. I was transferred from Valley Forge to Martin Army Hospital at Fort Benning. I got a surprise visit one day from Sergeant Wozniak. I couldn't remember anything that happened after I was hit, but he explained it all to me. He told me it was you who found me, took the radio off my back, ripped my shirt off, staunched the wound, and called a 3/4 ton truck to take me to the helipad. You were still under sniper fire.

You didn't have to do that. You would have done your duty to just call a medic.

"I want you to know that after three surgeries, I will get all the use of my shoulder. And I want you to know you inspired me so much, I no longer use the 'N' word. I respect you too much to say that.

Wishing you the best,

SGT Bryan Rocatta

DeAndre didn't know which surprised him the most—that Rocatta admitted he was a racist or his confession.

He gave Cabot the rest of the day off, then walked out of Camp Holloway into Pleiku city to the church where Smithy, Lee Anne, Chel, and Anh were. He'd just as soon his driver not know about his Jarai connection.

"Hey there. Welcome to Pleiku," boomed Smithy as soon as he walked into the church building. Scurrying out from a room in the back, Lee Anne and Chel added their welcome. After a hug, Chel looked at Lee Anne as if asking permission for something, and upon getting it, planted a big wet one on DeAndre's lips.

The trio's move to the provincial capital had been uneventful. DeAndre told them he would have very little to do in his new job. In fact, he had nothing to do except stay in touch with the province chief and General Ngô. The mere mention of Ngô made Smithy curl his lips in disgust.

About the only thing he had to tell them was that Anh and his family were in the process of moving to Pleiku. Anh had wormed his way into everyone's hearts, so that was good news to all. "When he gets here, let's all be there to welcome him," said Lee Anne.

The next afternoon, Smithy took advantage of the luxury of using the telephone to call DeAndre and tell him Anh had come. Of course, that meant Lee Anne's wish that they all be on hand for Anh's arrival didn't work out, but no one really cared. Anh had come to the church as soon as he entered Pleiku. His parents had many friends in Pleiku, so they had plenty of help settling in.

After a half hour or so of chit chat, DeAndre announced he had some paperwork to do and needed to return to his office. He invited Anh along.

"I've thought about it and changed my mind. I'd like to keep Smithy and Lee Anne in the dark about all this. You and Chel can tell them a few weeks afterwards—or better, when the questions about me stop coming," said DeAndre.

"Weeks? Sure. But why? They're not going to rat you out."

"No, but Lee Anne will get all maternal and try to talk me out of it. Besides, if they are questioned about it, they won't lie—not even to protect me. If you tell them when all the investigations are over, it should be safe. Let's just do it."

"Okay," said Anh. "I will tell Chel your decision tomorrow, which I'm sure will delight her. Then I'll bring her here and we can start to plan the details."

The three conspirators met in DeAndre's office the next morning and the basic plan was laid out to Chel, who thought it was very workable.

"Let's not meet tomorrow," said DeAndre. "I need to see Mr. Vu, the Province Chief so I can find out where the closest Ruff Puff bunkers are. Chel can start coordinating with the local FULRO."

Ostensibly, DeAndre needed to check with the Province Chief to get the location of Ruff Puff bunkers that would be nearest the small American sector of the Camp Holloway bunker line.

In reality, Anh had already shown him one. It was close enough to the bunker line for the Americans to hear shooting, but out of their sight owing to a stand of trees. He'd tell Mr. Vu that he had never worked with or even seen any Ruff Puffs before, and he thought he needed to know at least something about them. As long as Mr. Vu knew it was DeAndre out near the Ruff Puffs, that would substantiate the story about him going there for an 'inspection.' To ensure the Americans knew beforehand, he would tell the Military Police on the gate where he was going. If they recorded that in their logbook, so much the better.

The following day, Chel returned with news that local FULRO leaders were fully supportive and ready to take action at the designated time. Anh had also inspected the surrounding area to ensure no ARVN or Ruff Puff units would interfere.

The day had finally arrived. Knowing this was a pivotal time in his life, DeAndre was a bundle of nerves. He might never see his mother again or enjoy a steak at his favorite joint back home. He would miss Pastor Willis and the little white church. He'd have to get used to sitting on the ground. He might never see television again. But he also knew he had met the love of his life. He smiled as he thought how corny that sounded, but knew it was also true.

"Sir, the jeep is ready if you are," came the voice of SP4 Cabot as he popped his head in the office door. "Thanks, Cabot," came the reply as he picked up his M-16. "Let's get it done."

Anh was already in the back seat. As soon as DeAndre was seated, Cabot shifted to low and sped off.

DeAndre informed the Military Police at the gate who he was and his position. They looked a little puzzled, then realized they knew the head of the advisors to be LTC Arredondo. It was no surprise to them that the colonel was on his way home—that was happening a lot. The captain told them of his mission and said he would be back within the hour. "I'll be just beyond that stand of trees. No problem." He returned their salute and noticed one MP go inside, hopefully to write in the logbook.

The jeep pulled up to where DeAndre had instructed the driver to stop. Getting out, he saw the small bunker off to the side of the road and just a bit further was a bend in the road. Slinging his rifle, he walked up to the bunker, peeked inside, and found it empty.

Yelling back to the jeep, he said, "Well, this is not good. There's nobody in the bunker. So much for protecting the entrance to the village." With that, he turned to walk further down the road, but walking slower than before. He was still within Anh and Cabot's sight when three men, wearing raggedy uniforms, jumped out of the trees and grabbed him. While trying to resist, a fourth man, carrying an M-16, brought down the butt of the rifle hard on DeAndre's head, whose knees buckled.

Both Cabot and Anh had seen the incident. The jeep leapt forward but was met with automatic weapons fire coming out of the woods. A couple of rounds hit the jeep, which Anh thought was a bit too realistic. Cabot only speeded up. Another weapon up the road had a straight on shot of the jeep as it came forward. A full automatic burst into the radiator stopped Cabot from going any further. He and Anh both dismounted, and using the jeep for a shield, Cabot returned fire. The "enemy" fire stopped soon after the jeep halted.

Cabot and Anh crept forward and saw a limp DeAndre being hauled into the woods. Knowing they had only one weapon between them, they quickly agreed they couldn't fight multiple enemies who had automatic weapons. The best thing they could do was to haul ass back to the gate and tell the MPs. It was a quarter of a mile or so and they were both winded by the time they neared the gate.

Not just the MPs were alert, but the nearby bunkers also had soldiers running into position. Stopping outside the gate, Cabot breathlessly yelled as loud as he could, "Don't shoot! There's an American officer out there. We saw him captured. You might hit him."

The MPs quickly joined in the chorus, though everyone kept alert.

Ahn and Cabot entered the gate and within a very few minutes, were surrounded by senior level officers wanting to know the story. Cabot, of course, didn't have to remember any story and was forthright in his recollection. He even went so far as to tell his interrogators that his new boss was a little too gung-ho for him.

The officers gave Anh a harder time, trying to figure out why a Vietnamese civilian had gone out the gate with him. Anh only had to embellish the truth a little bit to assure them that as the son of a village mayor who had just moved to Pleiku, he was introducing his friend DeAndre to Ruff Puffs because he had never seen them before. That seemed plausible, and it looked as if everyone had swallowed the story.

With no infantry units on the post, it took a while to assemble a small force of security police to go looking for the captured officer.

CHAPTER THIRTEEN

They'd gone only a short way into the forest when the kidnappers released DeAndre, who smiled and thanked them in Jarai, though he did rub the back of his neck and head where he had been butt stroked. Surprised at his knowledge of Jarai, the men erupted in a chorus of words. It took DeAndre awhile to get them to understand that his speaking ability was limited.

The men quickly figured out how to slow down their speech and talk in only a few words at a time, and the communication improved immensely. They got him to understand that they were taking him to a village about four hours away.

Moving through the forest with greater speed than DeAndre thought possible, they also told him they had water and food for him. It turned out that "food" meant a few mouthfuls of rice, but it was sufficient to power him all the way to the village.

The hamlet of Plei Mrong was somewhat different than the Jarai villages DeAndre had seen. True, it still had a longhouse and had what he knew was a church building, but he also noticed the houses were styled a little different, with a wooden extension of the ridge carved into decorative patterns. The eaves of all the buildings had much more overhang than the ones he had seen. It also was larger than the Christian village he had visited with Chel.

Upon entering the village, they were approached by a man dressed in blue jeans and t-shirt with "I Love New York" emblazoned on it. He must have noticed DeAndre's interest because the first thing he said in passable English was, "Yes yes. I want go New York someday." With that, DeAndre met Ama Tum who was the village chief and the military leader of this contingent of FULRO. Reading the American's mind, the chieftain said he had learned English from the Special Forces soldiers who had trained and equipped the mountaineers.

Knowing that the captain's military uniform would not do, Ama Tum first asked DeAndre if he wanted some sort of loin cloth to wear. He also dispatched two men to return to Pleiku and make contact with Chel, asking her to send civilian American clothes to the village. Hearing Ama Tum's instructions, DeAndre also told them to ask Anh for the clothes and told them where to look. He didn't have many 'civvies' and he knew they wouldn't last long out in the bush, but they would have to do until he figured out a way to stay clothed.

The men returned just before nightfall, telling the big American that Chel would send the clothes out as soon as she could.

Food was laid in front of DeAndre. It was good basic fare of rice; some boiled squash, beans with hot peppers mixed in to give it zip, and a small piece of meat that looked like pork. It was quite good and he ate with gusto. He was glad that Chel had told him beforehand that the

Montagnards ate with their hands from a common bowl, family style.

But he had to shuffle his legs between having them splayed out in front of him or doubled up and sitting on them. Either way quickly became uncomfortable. Ama Tum noticed and smiled, "You like American Special Forces I know. Not used to sitting like Dega."

DeAndre had heard Smithy use the term "Dega" before and had asked the meaning.

"Dega means all highlands people. Not just Jarai or Rhade or Koho or Katu or Bahnar but all the hill people." DeAndre was also curious to know if all FULRO were Christian. "No, but most are," answered Smithy, "some Catholic, most Protestant. There are very few spirit believers in FULRO."

As he ate, DeAndre continued his conversation with Ama Tum, wondering what they expected of him.

"The American Special Forces taught us how to patrol, fight, set up defenses and all the soldier skills. But we have no idea how to organize ourselves or set up commanders. Maybe you help."

"Yes, I can help with that, but I will need to be sure that what I teach is something Dega can understand and work with. I do not want all of you to become Americans. Can you help me do that, Ama Tum?"

This brought a big smile, "I think you want to say you want to help in ways that fit Dega culture. Yes, we work together. We like that."

CHAPTER FOURTEEN

Mrs. Washington heard a car pull into her driveway that Sunday afternoon. She'd only been home from church for an hour or so and was just finishing her lunch. Peeking through the sheer window drapes, she saw two men in Army uniforms exit an olive drab Chevrolet sedan, accompanied by Pastor Willis. She knew they were going to tell her that her oldest son had died in Vietnam.

Already weeping when they identified themselves as Chaplain Curtis, Major Morales, both from nearby Ft. Stewart, Pastor Willis spoke next. "We have not come to tell you DeAndre is dead." He paused to let that sink in before continuing. "We don't know for sure, but we don't think he was wounded either but was captured by Communist forces and taken prisoner."

Stunned, Mabel Washington looked for a chair to collapse into. The Chaplain caught her arm and steadied her as she sat down. "He's a prisoner? Do we know if he is okay?"

Major Morales spoke next, "We don't know for sure that he is okay, but two people saw him as he was captured and pulled into the jungle by what appeared to be Viet Cong. He was walking and we know he wasn't shot, so we assume the best, but right now, there is no way we can find out. There are patrols out looking for him and the men who were with him are being questioned."

"What they're saying, Mabel, is that DeAndre is a hero but he's not dead," intoned Pastor Willis. "I think we can all celebrate that and know it is entirely possible we will see him sometime in the future. We just don't know when."

"I'm so happy to see you are in the care of such a good man as Pastor Willis," said Chaplain Curtis. "He tells me your faith is strong and that your congregation is a group of caring, prayerful Christians."

CHAPTER FIFTEEN

Two days after the "abduction," two Jarai came to the church in Pleiku. They had heard of Smithy and Lee Anne and were not surprised to hear a white man speak their language. Though Smithy had never been in the military, he knew how to practice operational security. After chatting with the men for a while and hearing what they said about FULRO and their village, he asked them for some kind of proof that what they said was true.

They handed him a folded piece of paper. Smithy was delighted and surprised to recognize DeAndre's handwriting, asking Chel to send his civilian clothes to the village.

When Lee Anne found Chel and told her of the Jarai men's request, Chel's peals of joy filled the air as she jumped up and down with happiness. She hugged Lee Anne mightily. Running to the church building, she took the paper from Smithy, but she didn't know how to read yet. All she cared about is that her beloved had touched that paper.

The men told her of DeAndre's request. She quickly told them that she would do more than send his clothes, she would deliver them in person. But first, she told them to stay the night in Pleiku.

Chel had another plan. She would figure out how to make more clothes for DeAndre before sending them out. Doing that, she would return to Pleiku after delivering the civvies. But first, she had to ask Lee Anne for help in making clothes. She knew how to make them by hand; she also knew Lee Anne had a sewing machine. Would Lee Anne help with making more clothes?

"Of course I will, Chel. Of course." was Lee Anne's reply. "Bring his clothes here and we will measure them."

Retrieval of his clothes brought Smithy into the picture. Cabot would certainly not let Chel have the clothes but would allow the American missionary to take them.

Smithy went to see Cabot the next day, and the clothes were in Lee Anne's sewing room soon thereafter. Lee Anne showed Chel how to measure them with a tape measure.

Lee Anne trundled out some cloth from her stash and soon Chel was sitting at the sewing machine with Lee Anne leaning over her. It was a treadle machine, so she had to both move the big foot pedal up and down to power the machine but also learn how to move the cloth under the needle. To start, they just sewed on a small piece of fabric, then graduated to learning how to cut out the sections before sewing them together. Lee Anne was not surprised at Chels' adroitness. She smiled inwardly and thought that DeAndre had made a very good choice in this woman.

As the two worked, Smithy called in the two Jarai messengers, gave them DeAndre's clothes and sent them on their way back, along with the message that more would be forthcoming.

CHAPTER SIXTEEN

Ama Tum and DeAndre began to assess his role as FULRO's newest fighter. They agreed they needed to be the aggressor and get out ahead of the attacks from Kinh troops. Their best tactics would be to ambush the ARVN soldiers traveling in their trucks. But, to do that, they needed good intelligence. That would require a spy. DeAndre knew just the man.

It was decided that DeAndre and a Jarai fighter would travel to Doan Ket and meet with Anh, but they found out he was now living in Pleiku. With that, the two went to Pleiku. DeAndre stayed well-hidden out in the jungle while the FULRO man visited Smithy's church, found out where Anh lived and talked to him. It was arranged for DeAndre and Anh to meet in Doan Ket three days later, simply because Anh could travel openly on the road while the other two had to walk through the jungle.

At the appointed time, the Jarai fighter met Anh outside the local primary school, then led him into the jungle to meet DeAndre. Anh had brought more clothes, and the big black man was delighted to see that the two women had sent not only typical western shirts and pants, but also shirts made of Jarai patterns. Not being keen on wearing a loincloth, he now had enough pants to obviate any need to do that.

The two man-hugged, followed by some quick catch-up chit chat.

"Anh, do you still go out to the Ranger compound?"

"Just to drink beer with a couple of the officers on Hoang's staff. Why?"

DeAndre had to take a moment to work up his courage to ask his friend. His voice tight from anxiety, he said, "Because we would like you to spy on the Rangers and let us know what their plans are. When the time and place is appropriate, we will ambush one of their convoys."

Anh was shocked. His mouth hung slack for a minute before he finally said, "DeAndre, if they figure out I ratted on them, they'll kill me."

"No doubt. I've never been a spy, so I can't give you any advice on being one. I'm just asking you so we can start to hit them. Yes, yes, I know. Our goal is to rid the mountains of all Kinh, and that would include you, so you are more than free to refuse. I wish there was a way to spy on other ARVN units so it would be much harder to link you to the ambushes, but I can't think of anything."

"First of all, FULRO can't execute an ambush every time I send information. It wouldn't take long for Thiếu tá Hoang to establish a pattern. In other words, mix it up. It would also help if you had another spy giving you information very different than mine."

The answer to that dilemma came soon after Anh had agreed to spy. He himself had an idea to expand the spy network. He started by including Smithy in the plot. He brought him up to snuff about DeAndre's situation and intentions.

Smithy had come to Vietnam to spread the gospel, not take sides in a racial conflict. He never did anything without prayer and consultation with his bride. But, after listening to Anh, he became inspired by DeAndre's bravery and devotion to the Dega cause. Both believed DeAndre's love and devotion to Chel was undeniable. Lee Anne agreed with her husband, as long as it did not interfere with their mission of ministering to the Jarai. He confided his basic principle to Anh: they must stay true to God's calling, yet do what they could for DeAndre.

"Tell you what I'm gonna do. I'm going to introduce you two to a lady who might just be of help." With that, he told Smithy and Lee Anne to dress up the next evening. Anh took Smithy and Lee Anne to a very nice, upscale (for Pleiku) social club. After they had ordered, he took the couple on a tour of the restaurant, introducing them to the local glitterati. One of them was an attractive and wealthy Kinh socialite who spoke excellent English, with just a touch of a French accent. Noticing that she was dining by herself, Anh asked her if she would like to join the three of them at their table.

Assuming Anh was romantically interested in the lady, Lee Anne readily agreed. As Anh seated her at the table, he introduced the three of them. Noticing the western style wedding ring on her hand, and listening to Anh's introduction, the Smiths quickly learned the two had studied together at UCLA. They were old friends. Her name was Pham Thi Thuy, married to an Australian. The two lived in Bien Hoa where they owned a prosperous tea and coffee export company in partnership with her parents.

She not only despised Ngô but had slowly become a discreet fan of the Dega. She had heard of DeAndre and she quickly realized she was sitting with three people who knew him well. She had heard of DeAndre's run-ins with Thiếu tá Hoang and asked if there was some way to champion his cause. Anh, with some input from Smithy, obliquely described how DeAndre was in need of intelligence about ARVN movements.

By swiveling his head and moving his eyes, he signaled that details would not be appropriate in the restaurant. There were too many ears. Smithy invited her to the church, and she smilingly said she would do so the next day. From then on, the dinner talk changed topics to Saigon politics and the state of the war effort.

Late the next morning, she arrived just as Smithy was telling a work detail how the walls of the newly reconstructed building needed to be painted. They walked outside and found Lee Anne working with a group of new mothers. After agreeing they would have more privacy inside the church, Lee Anne sent one of the older children to find Anh.

"Thank you for coming over, em Thuy," said Anh, using the western name order.

Offering the usual courtesies for a visit, Lee Anne offered coffee and tea. While making a pot of coffee, she glanced at the new acquaintance, noticing she was quite attractive, with a dusky complexion. Combining the strength of a western woman with the elegant grace of an

Asian, Lee Anne noticed an innate intelligence in Thuy's small talk.

"In case you are wondering, I learned my English while a student at UCLA. As I said last night, that's where Anh and I met. The French accent you probably noticed came from my time as a graduate student at Université Laval in Quebec.

"My parents live in Bien Hoa where they own a tea and coffee export company in partnership with my husband and me. My father is Australian, having lived in Ban Me Thuot during the mid 60s and my mother is half Kinh and half Rhade. I'm kind of a mutt." she said with a smirk.

She continued, "I met my husband after I returned from Canada. He was doing business with my father. Both my father and husband are thoroughly Australian."

If Smithy was taken aback by her directness, he didn't show it. "For such an urban sophisticate, I'm curious, Thuy. What brings you to beautiful downtown Pleiku?"

That tack broke the ice. Smiling, she told him that she often came to Pleiku to buy tea and coffee for export. Her frequent visits eventually led her to buy a small home. She shared a larger home in Bien Hoa with her husband.

"Smithy, there is a reason I told you about my lineage. I am very sympathetic to the Dega cause. Frankly, I think it would be great if they were able to establish an independent country. I would much rather deal with the growers than the Kinh. In other words, I am a huge fan of FULRO. I keep it quiet simply for my own safety and for my business. Anh only knows my feelings because we are old friends."

Smithy and Anh had dark thoughts that she might be some sort of femme fatale who was working for the South Vietnamese government. They both had ways of finding out. In the following days, Smithy talked to Dr. Pat, who had moved her charitable hospital from Kon Tum to Pleiku. She confirmed that Thuy was a straight shooter, having donated to the hospital numerous times and occasionally visited.

Anh talked to his father, who immediately misunderstood Anh's intentions and warned him that she was a married woman. With a laugh, Anh told his father he had known Thuy way back in college and had even met her husband. Relieved, the father opined that Thuy was not a southern spy.

After these confirmations, Smithy and Anh asked her to come to Dr. Pat's hospital where they laid it out for her—DeAndre needed help with intelligence gathering. They hadn't even finished their explanation when Thuy interjected, "Of course I can help. It is not unusual for me to socialize with army officers, both at the social club and some of the restaurants in town. I am sure I can dig out a little helpful information once they have had enough to drink," she said with a slightly lusty smile. Even Lee Anne understood what she meant by that. There was no doubt Thuy could flirt with great skill.

But they all agreed there needed to be a secure way to transmit information without seeing each other and without specific detail. It was decided she could use one of the many couriers who carried business information for her. They were in and out of her office several times a day, usually carrying an envelope with letters about buying tea. To avoid putting too much information in one letter, there might be several letters, some with fake data with a short code that meant it was fake. It was the perfect cover. They would take the messages to Anh's home where even his father would not question comings and goings.

It took a while before the spying paid off. While at the social club one night, a semi-drunk ARVN officer had invited her to join him at his table. She had seen him there on numerous occasions and knew he was a lower ranking member of General Ngô Du's staff. Surely a II Corps headquarters lieutenant would know of troop movements.

As she sat down at the table, she leaned over the young Trung uy in such a way as to give him a good peek down the front of her dress, all the while chatting amicably. She had removed her wedding ring before entering the club and the young First Lieutenant thought she was available. It didn't take Thuy long to convince him she was not only available, but ready. She waived over another round of drinks. Making sure he got more scotch whiskey, she ordered her "usual." She had arranged with the bartender that her "usual" was to be plain ginger ale.

"Ah, you're from Saigon," she began. "What do you think of the highlands?" Predictably, the city boy was not enamored with the mountains.

She gently led him in the direction of talking about the Jarai and other tribes. "You mean the stinking moi? Geez, I hate them. Hell, they can't speak Vietnamese, much less dress in a civilized manner." She knew now that she had the right man.

"Yeah, but you're on the staff of one of the most powerful people in the country," she went on. "I mean, because you are a Trung uy, you don't originate operations, but you help map them out for the ground troops. That must be complicated work—but I bet you are good at it."

She ordered another round of drinks. She stood to talk to some passing friends, then gave him another long peek while reseating herself.

It took about an hour before her efforts paid off. A regular ARVN battalion was to attack a Dega village named Thanh Ha. Intelligence had revealed that there were a number of northerners as well as a few local Viet Cong in the village whose job it was to protect a cache of weapons and ammunition. The young officer indicated that since the village was well away from the road, the tracks carrying the troops would have to leave their compound in Pleiku at an early hour the following Friday.

Thuy noticed he was slurring words badly by this time, so she ordered another round of drinks and excused herself to go to the ladies' room. When she returned to the table, the Trung uy was sound asleep with his head cradled in his arms.

By the end of the day after Thuy's conversation in the club, the information was in DeAndre's hands. He and Ama Tum agreed that this was the kind of intelligence they needed and began to plan the operation.

They knew where Thanh Ha was located and knew it would take about a day to travel the distance on foot through the jungle. They also knew that though the ARVN troops had superior firepower, they also were not used to being in the jungle and would not pursue the Dega. It was decided to ambush the Kinh troops just before they entered the village. That way, the enemy would not have time to organize a static defense inside the village. Being a long way from the road, giving the FULRO fighters much more time to leave the area before ARVN reinforcements arrived.

DeAndre was impressed with how well the fighters were organized into small teams of three each. He realized they had been well trained by the American Special Forces. Such teams allowed Ama Tum to concentrate the teams for maximum firepower, then quickly disperse, all the while staying in contact even though they had no modern communications such as radios.

The plan was excellent. The DEGA would allow the five-soldier vanguard of the ARVN into the village, fire a few rounds at them, and wait for the word to get back to the battalion headquarters. DeAndre and Ama Tum made a scientific wild-assed guess as to where the headquarters would be. Good reconnaissance had shown there was a well-worn path into the village and ARVN being ARVN, they would follow the path.

The day before the army was to attack, the FULRO men laid booby traps on the far side of where the ambushers would be located. The Kinh soldiers would get a hail of concentrated fire from the DEGA, who hoped the ARVN would fall back far enough for them to trip booby traps. The secret was to put heavy fire on the ARVN troops for a few minutes, then on Ama Tum sending a red flare into the sky, they would cease fire and melt back into the jungle. In other words, the ambush would be short and sweet and very violent.

On the morning when the fighters would leave their village for the trek to Thanh Ha, they had a prayer meeting. With raised hands and the laying on of hands, the families, children, and elderly prayed heartily for success and the safe return of their loved ones. DeAndre was reminded of the sendoff he had received at his mother's church.

It took fourteen hours for the team to walk through the dense jungle. DeAndre stuck close to Ama Tum as he marveled at the ability of the men to move quickly in the thick growth. They arrived after dark, ate a sparse meal, then slept. First light revealed they were close to the chosen ambush site. Teams went out to set the booby traps while the others prepared well-camouflaged positions to use during the firefight. They had to avoid being seen by the villagers because some of them might be inclined to advise the ARVN.

Knowing the southern troops were disembarking their trucks and beginning to move in their direction, Ama Tum had sent out a couple of listening posts to herald their arrival. Sure enough, just as hoped, the ARVN vanguard was spotted about mid-afternoon. They were allowed to pass through the ambush site, but when they stepped into the village itself, they were met by a few shots that dropped two of the ARVN soldiers and caused the remaining ones to use the radio to call back to the headquarters.

Ama Tum's knowledge of how the ARVN operated was perfect. When the vanguard was fired on, the commanding officer and his group were right where the FULRO fighters wanted them to be. The signal to fire would be a shot from Ama Tum. He took careful aim at a man walking in front of two radio operators, fired and watched the man falter and pitch forward. A split second after the first shot, two more rang out. Down went the two radio operators. The other soldiers began to wheel in the direction of the shots, only to be met by a fusillade of concentrated rifle fire and hand grenades.

More ARVN troops were seen to fall, while others sought cover behind them, thereby tripping booby traps. Their return fire was undisciplined and poorly aimed, most just firing wildly in the air. For five minutes, the FULRO men poured in fire, but when a red flare appeared over the ARVN, the ambush abruptly stopped.

As they moved back into the jungle, DeAndre knew they had inflicted a major hit on the enemy and reduced their own casualties to three lightly wounded and nobody killed. When they arrived back at Plei Mrong, they were ecstatic, as were their families.

Again, the village celebrated with a joyous prayer meeting. Afterwards, DeAndre felt a deep need for sleep.

But there was more joy to come for the village. Chel and DeAndre had talked to Smithy when he came out to visit members of his congregation in a village to the north of Pleiku. A decision was made and plans hatched.

A month after the ambush of the ARVN battalion at Thanh Ha, Smithy and Lee Anne came to Plei Mrong, bringing Chel with them. Chel had already talked to

Ama Tum and received permission to hold the wedding in the village prayer house.

Every one of the villagers wanted to be a part of the festivities. Besides the wedding itself, there would be a feast. Husbands and wives helped decorate God's house. Men selected two pigs to slaughter and figured out when and where to cook them. Children were tasked with finding flowers for the bride and the feast site. The women set up the dinner area with mats and palm fronds to sit on. It was a joyous day of happy busyness.

Smithy performed the actual ceremony but asked Ama Tum to participate as well. The congregation eagerly came forward to lay on hands and ask God's blessing on the union. An earsplitting roar ensued when Smithy told DeAndre he could kiss the bride and willingly endured a little good-natured teasing from Lee Anne that he may have taken a bit too long to kiss Chel.

The feast was all one could hope it to be. The pork and trimmings were delicious, accompanied by many Jarai words of congratulations. People spontaneously sprang into singing praise music. As the sun grew closer to the horizon, Ama Tum took the couple into the long house and showed them their room for the night.

They both noted that they would not be sleeping on mats on the floor, but rather on a raised platform with a soft plaited mat suspended between the wall and two poles. But there was also a blanket for them, which could be used for both warmth and cushioning. After exchanging shy smiles, they went back outside to rejoin the crowd.

DeAndre wasn't sure when the feast would end or

when he and Chel should take their departure and go to their room. As they held hands, he didn't know that Chel was equally apprehensive. She had the same questions, but she was also nervous about the night. She was not a virgin, but was that important to DeAndre? Would he be able to detect that she was experienced? She had talked to Smithy and Lee Anne enough to know that America was not a matriarchal culture, and while she was ready to lustily initiate the love making but was worried that might offend her new husband. Neither of them was nervous about the sex per se, but both wanted to be sure they didn't offend their new mate.

One of their mutual questions was answered when Ama Tum approached them, saying quietly and delicately that it would be okay for them to leave the feast whenever they wished. It really was up to them. If they wanted to leave and go to the room, all they had to do was get up and walk away, without the need to make any comments or excuses to the villagers. They both spoke at the same time, thanking Ama Tum.

Chel was both eager and hesitant. She desperately wanted to show DeAndre how very much she loved him. Finally, her desires overcame her nervousness and she squeezed his hand, then nodded towards the long house. Without a word, they arose together and went to their room, not saying anything to anybody.

The door closed, they kissed—a long and passionate kiss. Chel lost her nervousness when she felt the bulge in his trousers. Breaking away and stepping back a little, she ran her hands down his face, feeling and caressing his eyes, nose, and mouth. Then she slowly continued down his chest, and finally, to his groin. Looking into his eyes, she saw the lids close slowly as he enjoyed her touch. Neither cared about the other's culture any longer. They just wanted each other.

Chel began by unbuttoning DeAndre's western style shirt with slow and deliberate moves. He reached towards her to begin disrobing her, she gently brushed his hands away while continuing to open his shirt. He obediently let her have her way. When she opened his shirt, she noted that he had hair on his chest, unlike Dega men. She took her time to explore the feel of male chest hair. Making it easier for her, DeAndre took his shirt off, leaving it on the floor.

Clasping both her hands in his, he gently moved them away from his chest and guided them to her sides. With a gentle pat, he indicated he wanted her hands to stay there. His hands shaking ever so little, he began to unfasten the beautiful Jarai shirt she wore. He wanted to go slow. Knowing since he'd met her that she had larger breasts than most Montagnard women, he wanted to prolong the enjoyment the first time he held them in his hands.

Even before he removed her top, he reached in to caress one breast, gently kneading the nipple. Now it was his turn to watch her eyelids close in lusty enjoyment. Soon, he removed her shirt, unashamedly admiring her chest. Slowly bending over, he gently kissed one nipple, then the other.

That was more than Chel could stand. With a bit of a shudder, she stepped back and began unfastening his belt.

No longer wanting to go slowly, she had him step out of his pants, revealing his very erect manhood. Before DeAndre could object, she quickly finished undressing, then gently kissed it. Her hands on his hips, she guided him backwards until he felt the bed mat on his legs.

With a motion that he should just stay still for a moment, she moved to the side of the bed, unfolded the blanket, and placed it where they would lay, with an extra fold about hip level. Taking his arm, she guided him to lay on his back.

Bowing over him she bestowed a long lingering kiss on his mouth, then slowly lowered herself onto him. Once she was completely penetrated, they lay still in each other's arms awhile, both enjoying the moment and anticipating what would soon happen. Finally, she propped herself up, seized both his wrists, pinning him to the mat. Chel proved to be a tease. He couldn't move his hands, and she brushed her nipples across his mouth. Finally, she stopped and placed one nipple in his mouth, holding still. He sucked on her, but also unexpectedly flicked his tongue on the nipple. That was a new sensation for her, and she moaned in ecstasy.

Releasing his wrists, she sat up a little and began to move her hips slowly. Both groaning with rapture, he reached up and cupped her breasts. Chel's hips moved faster. She looked down at their coupling and moved up and down on him. The delicious agony for both was palpable. As her pumping bordered on frantic, she suddenly arched her back higher, allowing DeAndre to do the thrusting. Both exploded in their orgasm's, letting loose groans that the people outside must have heard. Neither cared.

As the thrusting subsided, she lowered herself to kiss him in grateful thanks for this physical way of expressing their deep love for each other. She slid off him and moved to his side, putting her head high on his chest. After murmuring, "I love you" to each other, they drifted off into a wonderful sleep.

There are no clocks in longhouses, so DeAndre had no idea what time it was, but he awoke with a need to relieve himself. Coming back to bed, he looked down at his gorgeous wife, admiring her body and her smile and sensuous demeanor. His own body reacted. He gently reached down to kiss her breasts, eliciting a groan from Chel. He moved to kiss her mouth as his hand slid down to her groin, found her opening, and began to massage her trigger point very gently.

As her excitement rose, Chel didn't open her eyes, but her face slowly began to contort as her delicious agony increased. DeAndre became more excited as he watched his wife about to explode. With one quick but strong motion, she rolled on her side and reached across his body to pull him on top of her. Not a word was said as he entered her, and they finished the journey together. Their ecstasy complete, they again drifted back into sleep.

He awoke when he saw there was daylight in the room and he heard people talking outside. His stirring woke her and she yawned herself to life. After sharing one more kiss, they dressed and walked outside onto the longhouse's porch and were welcomed by the villagers.

One woman beckoned them to come to her area where she served them both breakfast. With a smile, she spoke to Chel who laughed, then translated for DeAndre. "She told me not to expect this every day."

They had finished about half of their food when Ama Tum approached with a grim and hurried look on his face. He attempted to speak in English, but DeAndre stopped him and told him to use Jarai instead. That worked better because whatever part DeAndre missed, Chel would translate for him.

Their honeymoon would be short. Anh had sent a courier to Plei Mrong with information about an impending Ranger raid.

CHAPTER SEVENTEEN

Thuy's beer drinking with Ranger staff officers again paid off. Thiếu tá Hoang would be leading another of his infamous raids on Jarai villages. Of course, after the Eastertide campaign when Kon Tum almost fell, there were no more Kinh in the villages.

DeAndre now knew that General Ngô himself had decreed that FULRO was a greater enemy of South Vietnam than the Communists. Both Ama Tum and DeAndre knew the road leading to the village, and DeAndre understood immediately that one of the big curves would be a great place for an ambush. Sketching out his plan as he went, Ama Tum was impressed with DeAndre's idea. They had five days to prepare, and one of those days would be walking through the jungle to the ambush site. That didn't leave a lot of time.

But Ama Tum had one point he insisted on—that DeAndre not participate in the firefight. If Thiếu tá Hoang saw or heard about the presence of the big black man, it would greatly increase the amount of pressure on the FULRO movement. General Ngô would love to capture DeAndre and show that Americans were now leading FULRO. DeAndre reluctantly agreed. He saw Ama Tu's point, but that didn't assuage the feeling that he was not doing his job fighting the ARVN.

The FULRO fighters knew that most of the time, an anti-mine team preceded any Ranger road movement. They also knew the mine sweeping team moved well ahead of the Rangers. There would be a twenty-minute window to place any explosives in the road. DeAndre was familiar with the road and the ambush location. It was the very place that had made him nervous when he first joined the advisor team on his first mission with Hoang. It would take about three hours for the trucks to travel from their camp at Kon Tum to the ambush site.

The day before the ambush, the FULRO fighters arrived in the area, but away from the road a bit. The leaders of the various teams involved in the ambush carefully reconned the location for their teams, including the group that would be at the curve and would have a straight shot down the entire length of the convoy. Other team leaders found their areas along the side of the road, noting places for good concealment as well as quick escape routes. The explosives team looked for the best place to plant their device, trying to find a place that would have maximum impact and be a place where digging in the road would be the least noticeable.

The entire force then set up a night encampment. Everyone checked and rechecked equipment and ammunition. Team leaders again reviewed the role their group would have. The nervousness was palpable, as it is with any group of soldiers who know combat is close. Each fighter had brought extra rice. There would be no breakfast the morning of the ambush, but rather all would eat a handful of rice while moving away from the area after the fight.

DeAndre had been successful in convincing Ama Tum to change one important tactic. It was essential to limit FULRO casualties.

Even before dawn broke, all were awake and checking gear again. Ama Tum had a quick meeting with all the team leaders to review the signals to open fire and for disengagement. When that was finished, he led them in prayer, asking God for His help

As first light broke, the fighters moved towards the site, each team under the guidance of their leader who knew the way and knew the ambush site. As always, DeAndre was impressed by both the speed and the quiet of moving through the jungle. They needed no machetes nor did they get entangled in the "wait a minute" vines.

About an hour later, they arrived. Working with their men, the team leaders found well-camouflaged places for each, being sure they weren't too far apart. The team in the curve itself was the largest team as they had to bring maximum violence on the Rangers. The smaller teams strung alongside the road would wait until the Rangers started to leave their trucks and run for cover before opening fire. It would be a straight on shot as the Rangers blindly headed for what they thought would be cover. Within an hour, all was ready. All that was needed was for the ARVN mine sweeping team to pass by.

Reluctantly, DeAndre took his place out of sight of the road.

The Dega were a very well-disciplined group. There was no moving around trying to find a more comfortable position. There wasn't any swatting insects. There was no talking. There was no scratching. There was only an alert watchfulness waiting for the enemy.

Forty-five minutes later, the tension heightened as the ARVN mine sweeping team appeared. DeAndre was surprised at the carelessness of the sweepers. None were on foot, and only one metal detector was deployed, and that very carelessly dangled outside the jeep by the operator. He hoped the Rangers had also become less cautious.

The team went by and the explosives team went to work, scrambling onto the road to a preselected place for planting the unexploded 105mm American artillery round. They had practiced this numerous times back in the village and they had it in place within eight minutes. Only a really observant person would see where the soil had been disturbed in the road.

Another ten minutes went by before Thiếu tá Hoang's jeep appeared, followed by a column of American-made two and a half ton trucks. Ama Tum had the detonator in his hand, waiting for the precise moment when the jeep would be atop the explosive.

The explosion hurled the jeep into the air, throwing bodies in all directions. At the same instant, the team in the curve opened up, raking the sides of the trucks on the column, hitting many of the Rangers before they really understood what was going on. Though they avoided using full automatic on their rifles, the FULRO fighters kept up a high volume of fire that was also well aimed.

As it dawned on the Rangers that they were being attacked, they began to jump over the sides of the trucks, seeking cover in the roadside brush. Those who bailed on the left side were fine. Those on the right were met by a cacophony of bullets coming from invisible places in the bushes.

Within seconds, there were Rangers stacked on top of other Rangers, some dead and some wounded as they lay on the road. The fire continued for about two minutes before Ama Tum signaled the team in the curve to cease fire. The roadside teams took their cue from the main team and also ceased fire. They did not want to expend all their ammunition in case they had to fight later in the day.

The tactic DeAndre had insisted Ama Tum adopt was not to stay until every Ranger was dead. He knew that even though the command jeep was out of commission, somewhere in the column of trucks was another radio and some officer would be calling back to higher headquarters for help. Before too much longer, there would be artillery impacting in and around the road, followed soon after by helicopter gunships. The rebel fighters made a hasty and orderly withdrawal to their assembly area.

Ama Tum and the men were overjoyed. They had suffered only one casualty, and he suffered only a minor wound. The men knew they had inflicted a major loss on the Kinh. But now, they wanted to return to the village as quickly as possible just in case the Vietnamese were smart enough to know the home base of their attackers.

The daylight disappeared as the fighters neared their homes and were properly welcomed as heroes. Lamps fueled by chunks of slow-burning tree resin gave light to the few shared tattered Bibles and hymnals as Christian praise songs echoed through the otherwise uninhabited forest. Soon afterwards, there was a worship service held at the chapel. DeAndre again thought of the Sunday service at his mother's church during which everyone prayed that he return home safely. He wouldn't be going back to his American home, but he was safe—for now.

Chel stood next to him, excited to have her man back with her. Holding his hand, she kept looking up at him. A smile and a squeeze of her hand and he knew what the night foretold.

Ama Tum and the village leaders had decided to make the room Chel and DeAndre had used on their wedding night a gift to the couple and let it be theirs. In the next few days, Chel set about turning the room into a proper home. Some of the men worked with DeAndre to build a granary near the entrance to the room as well as build a small chicken coop on the outside wall. He found a place to keep his small store of clothes, Chel found some cooking utensils, he hung the M-16 that had been assigned to him on the wall, and Chel found some warmer covers for their bed. She also got word back to Lee Anne that she would like the sewing machine sent out to her.

CHAPTER EIGHTEEN

There had been one major shortcoming to the ambush on the Rangers. Normally, the men have a chance to take rifles, ammunition, and grenades from dead Kinh soldiers. Though DeAndre's tactic of moving out of the area before the Rangers had a chance to amass overwhelming firepower had been successful, it had also meant they not only had not gained any rifles but had not replenished their ammunition either.

The village leaders talked about the munition's situation in particular. They needed more M-16 and AK-47 rounds. Someone said that the southern government was once again trying to move Kinh into new villages in the mountains. What was different was the emergence of FULRO so the Kinh villages were patterned after the fortified villages used by the Diem regime back in the early 1960s. That meant there would be weapons and ammunition. The closest of these was three days away. It was not in the jungle but rather in a large clear area which would not give the Jarai any cover.

DeAndre's leadership rose to the top again. The Jarai had never fought in any other place than in the jungle and had no idea what tactics to use to attack a fortified village that was built in the open. He started by suggesting the attack take place during the early afternoon siesta. He remembered well the first time he had been with the Rangers when they circled their trucks, slung hammocks, and napped after eating lunch. He knew Vietnamese of all stripes were used to napping then and hardly suspected anybody would attack them during that time.

The Dega were not as familiar with the ways of Kinh as DeAndre, and it took him some time to convince him that this would be a good time.

While the village was not in the jungle, the surrounding terrain was rolling hills. By keeping in the lower places, small teams would be hard to spot advancing on the village.

The FULRO leaders spent the entire next day developing their battle plan. There would be blocking forces hidden in the tree line on the opposite side of the attack. The main force would be organized in small teams of two or three fighters. These would be less noticeable than a massed force. Some teams were given the mission of capturing any Kinh seen outside the village, such as those tending the kilns used to make charcoal for cooking.

Other teams were assigned the task of retrieving weapons and ammunition from dead Kinh as well as any found in houses and stores. It was also decided that they would not open fire until the enemy fired the first shot, further hiding the attack. Finally, as many members as possible of the main attacking force would be clothed in western clothes to help conceal that they were indeed Montagnards.

The evening prayer time was filled with many pleadings to God to help them in attacking the village. The women now learned of the plans, many of them openly nervous about losing their husbands. Their entreaties were to save their men. Chel was among the wives who prayed openly for success and the safe return of her man.

After another day of planning, the FULRO fighters left Plei Mrong, laden with food for the week-long mission. They packed their western clothes as it was much easier to navigate the jungle in their native loin cloths. The night before departure, Chel made sure DeAndre would leave with a smile on his face.

The journey was uneventful. Arriving at the edge of the wood line surrounding the village's field, the men were able to fine-tune their mission according to what they saw. After everyone had changed into western clothes, DeAndre led the blocking teams as they split from the others. Keeping within the forest, they made their way to the back of the village to be in position the next day. Once again, to his increasing dislike, DeAndre was to be kept in hiding in the trees. The main group split into their various teams, each going to a position in the wood line nearest to their assault position.

The night was uneventful, as was the morning. The tension became palpable. As the siesta time began, each team left their hiding place at a different time to reduce the chance of being spotted as an attacking force. Their reconnaissance had revealed two charcoal kilns, and the assigned teams headed towards those. While it was impossible to hide their weapons, they did try to minimize the chance of them being seen by slinging their M-16s and AK-47s in such a way that the rifles were on their backs.

The two teams assigned to the kilns found villagers there, but they were sound asleep and taken prisoner without problem. It wasn't until the Montagnards were within the outskirts of the village that they were spotted and the alarm raised. The local Ruff Puff team charged with protecting the hamlet were jolted awake by the yelling, but it took them a while to see the groups of Moi soldiers. Only then did any shooting begin.

The shooting spurred the inhabitants to dive for bunkers and trenches.

It was not unexpected, given the poor reputation of Ruff Puff units, but it didn't take long for DeAndre's blocking force to see action. As the terrified Kinh soldiers ran towards the wood line, they were cut down by the well-aimed shots of the Dega fighters. The civilians within the town were herded into large groups where they could be checked for weapons.

No Ruff Puffs were found inside the village. All of them had tried to escape, only to be cut down. DeAndre noted that the FULRO men had no intention of taking prisoners. Though that was against his own personal morals, he also understood the security risk of moving prisoners to Plei Mrong. Besides, what would the Jarai do with them?

Both the main assault force and the blocking teams converged on the killing field, scooping up Kinh weapons and ammunition. Once finished, they all moved into the jungle. Rather than going straight home, they made for a spot they knew they could defend easily, yet would be hard for the searching ARVN to find them.

Again, a successful raid. One man had been wounded, and that was nothing more than a scratch on his lower leg. Another fighter had been hurt worse when he fell into a trench and banged a knee on a villager's shovel.

They all knew it would take three days to get home, but besides wanting to see their families, they worried the Rangers might raid Plei Mrong in their absence. They decided to cut the time to two days, moving at night.

The unexpected arrival of the men back home was a thrill for the families, happy again that there had been no deaths. After sending a runner to Pleiku to thank both Anh and Thuy for the intelligence, DeAndre joined his wife in their room.

Lying in each other's arms later that night, DeAndre explained the hard facts of being a soldier to Chel. He had been in enough combat to know that two firefights with no fatalities was highly unusual, and he wanted his wife to understand that sooner or later, some men would not be coming home. It was true that some of the men had field medical training from the US Special Forces, but that didn't mean no one died.

But DeAndre worried. He was uncomfortable to having families living with troops. He thought that having wives and children around would make them too cautious. He talked to Ama Tum the next day. He held an opposite viewpoint based purely on the fact that neither he nor any of the FULRO fighters had ever lived away from their families.

"Do you want to be away from Chel?" he asked. That question brought a smile to DeAndre's face. Of course he didn't want to be away from his wife. "I see what you mean, but also, when we come home from a fight, we try to be cautious and not leave a trail. But we might also get careless and leave a way for the ARVN to follow us." The two men were now friends, and they agreed to keep talking.

Remembering that to lessen the chances of being followed home, they had not returned directly home after their last attack. DeAndre used that as one of his arguments. One morning as the two men worked in the garden, they came to an agreement. They would continue having a village like Plei Mrong with families and homes, but they would also have a semi-permanent "battle camp" of just fighters. They could move the site of the camp frequently, making it more difficult for the ARVN to find the families and it would allow them to move quickly if intelligence indicated they had a short time frame to react.

They brought up the idea with the entire group of fighters, most of whom thought it a great idea. Several had some ideas where to locate the first battle camp. A small recon team was selected to visit three of the suggested locations. DeAndre was ostensibly the leader, but he didn't know his way around the jungle as the Dega did. His expertise was in site selection.

Back in Plei Mrong, it was decided where the first camp would go. Then, and only then, did the entire village, including the women, get together. The only objection they had was that there needed to be some way to relay information about a firefight back to the wives as soon as possible.

That seemed more than reasonable, and it was agreed to do that. The leader of a mission would decide how to contact the village in such a way as to reduce the chances of the messenger being followed. The women were very much in favor of the idea but also insisted on having a plan that would send the men back to Plei Mrong one at a time for a two-day visit.

Four days later, DeAndre set out with a third of the fighters. Their mission was to establish the camp, then prepare it for the remainder of the men. The site was to stay simple. No buildings, except for a palm frond wall that served as the front of the chapel. Some stocking of firewood was allowed, but the stacks kept small. Hammocks slung between trees was the preferred way to sleep. Very few trees were to be cut down, and then only small ones. A series of listening posts were to be hidden in a circle outside the camp.

Within a week, the rest of the men from Plei Mrong were also in the new camp. Not long after, they got word that new FULRO fighters would soon join them. The movement's leaders wanted to balance out the size of the different tribes' forces.

Within a week, new FULRO warriors joined them. They are from many different tribes, but all of them either spoke Rhade or were learning it. Back in the mid-60s, the leaders had decided that Rhade would be the lingua franca of FULRO, mainly because Rhade intellectual

Y'Bham Enuol was the founder and President of BAJARACA, the precursor to FULRO.

DeAndre settled in as the de facto leader of the camp. A Jarai man named Mip was ostensibly the boss, but he deferred to DeAndre, acting more as translator.

DeAndre also wanted Anh to know about the battle camp. Instructing a messenger, he said, "Tell him about the base, but also tell him that for good operational security, we will not let him know each time we move to a new location. The people at Plei Mrong will know."

They had a two-week period of adjustment as they tried to learn Rhade. They also had to learn how to cook for themselves, being used to their wives' feeding them. DeAndre's idea of having listening posts was a new idea to them, but it didn't take long for them to recognize that it was a good idea.

A runner came in from Pleiku and went to Mip to give him the latest intelligence. But before the courier began talking, they both met with DeAndre.

"A regular ARVN battalion will attack the Jarai village of Plei Doch within the week." Thuy's dining with the drunken ARVN lieutenant had paid off again.

Asking Mip if he knew where Plei Doch was, and getting a positive nod, De Andre told the messenger to continue. There wasn't much more intelligence except that the ARVN unit had just arrived in the mountains after having been in the coastal plains of Binh Dinh Province.

There was no exact date for the ARVN raid, only "within the week." Mip explained that the village was a long way from a road, meaning the Kinh troops would have a long march through jungle, and these soldiers would not be familiar with moving through the thick forest. DeAndre suggested the FULRO fighters hit the ARVN in the jungle rather than the road, simply so they could harvest more weapons and ammunition. After sending a fighter to Plei Mrong to inform the families, DeAndre called a meeting of the leaders.

"I want each of you to feel free to give ideas." This was a new approach to the men who were used to leaders just telling them what to do and not asking for input from them.

"It will take the ARVN about four hours to march from the road to Plei Doch—maybe longer since they are not used to the forest. I propose we set up an ambush located about two hours from the road. That way we can collect more weapons and ammo afterwards."

The men liked the idea, though one piped up and asked how the FULRO would know the path of the ARVN—in other words, how would they know where to set up the ambush.

Mip spoke up. "Maybe we put out spotter teams between the road and Plei Doch. That way we can tell their path." Others spoke up, suggesting that the ambush site would allow the FULRO fighters the chance to fine-tune the location at the last instant without being spotted by the ARVN. Others said they liked the idea of three-man teams they had used during their attack on the Kinh village.

"I think we have a good plan," said DeAndre. He asked for volunteers to recon the area and determine the ambush site. Another group under Mip's command set out to locate the spotter team placements. Both groups left well before sunset to allow the main body to leave the camp as soon as possible.

Two days after the arrival of the runner, the fighters moved out, leaving only a few teenage boys behind to protect the camp. As always, DeAndre was still amazed at how easily the Montagnards moved through the jungle. There was no slashing of a machete by the lead person. Nobody's rifle got tangled in the "wait a minute" vines. Noise was kept to a minimum, increasing security.

The main force was still settling into the ambush site when word came back that the ARVN troops were disembarking from their trucks. The spotter teams came in and were incorporated into the main body. The ARVN's path seemed that it was on the path necessary to bring it into the ambush's killing zone. Again, DeAndre was told to stay out of sight.

Perfect. The ambush was executed flawlessly. Mip opened fire, unleashing a barrage of small arms fire. When the ARVN tried to run in the opposite direction of the shooting, they were met by a network of Claymore mines that were detonated by FULRO fighters up in trees. Panicked southern soldiers tried to retreat by the same route they had taken from the road, only to be met with more fire. As Mip called for a cease fire, the shooting died down, with only the occasional shot by a FULRO fighter dispensing of a wounded Kinh soldier.

As the FULRO fighters scooped up weapons, DeAndre came out of hiding and looked for anything that might be actionable intelligence. DeAndre knew that 1st Lieutenants were company commanders in the South Vietnamese army. He spotted a dead ARVN Trung uy, and after checking to be sure the body had not been booby trapped, DeAndre began going through the man's backpack and pockets. He kept some printed material to be translated later.

Then he found a waterproof wallet consisting of material like Ziploc baggies that were bound together in a sort of book. It was a great way to keep photos and love letters from getting wet and were often used by American troops too. Leafing through the photos, he came upon a photo of the man in a dress uniform standing behind an older couple who were seated on a couch. Obviously, they were the officer's parents.

For a moment, the war ceased to exist for DeAndre. He pondered the family of the dead young man and the grief the parents would soon suffer. He thought more about the similarities between the lieutenant and himself. DeAndre still had his loving mother but knew his father had loved him as well. The lieutenant still had both parents. Maybe the Trung uy was single but maybe had a girl back home. DeAndre was a newlywed who had lived through horrific combat to marry the girl of his dreams. They both had followed the calling of their governments to fight a war. The only difference was that DeAndre was alive and the ARVN officer was dead. Why? What fate—what God—what karma had determined who lived and who died? The thought came to him that he would challenge his mother's pastor to answer the question: if there is a loving God, why are evil and war allowed to exist?

Back in reality, weapons and ammunition were scooped up and the FULRO fighters moved out of the area swiftly, making a side trip to nowhere on the way back to camp, just in case some ARVN were tailing them.

Back in camp, some of the men were given passes to visit Plei Mrong for two days. It was DeAndre's turn to do the cooking for the camp. Flipping and turning squash, cabbage, some peppers and rice together in a big wok, he became aware that there seemed to be a lot of air activity. Curious, he noticed a lot of fixed wing aircraft that seemed to be taking off and landing at Pleiku.

"Mip. See all the airplanes?" Mip nodded his head affirmatively. "Maybe Americans leave Pleiku?" That thought had been in the back of DeAndre's head for a long time, but now it blossomed forth. No home cooking at the Air Force chow hall. No Smithy and Lee Anne. No American presence to tamp down General Ngô.

DeAndre realized he really was alone in the mountains of Vietnam. He could not go back to being an American again, even if he took Chel with him. He thought of his terrified mother who didn't know what had happened to him—and now never would know. Had he really made the right choice? He loved Chel with all his heart, mind and body, but doubt still crept into his soul.

Mip looked at his new-found friend, saw the look of deep thought on DeAndre's face, and decided to leave him alone.

Two of the men who had taken passes to go to Plei

Mrong returned that evening. Suspicion turned to hard facts—there were no Americans left in Pleiku. Trucks had loaded up American troops, while civilians bribed their way onto the few remaining buses leaving town.

It was quickly decided that the entire camp would return to Plei Mrong immediately. All the FULRO men and their families needed to talk about this. DeAndre longed for more information but had no idea how to get it.

He was greeted by Chel with a passionate kiss and a whisper of future delights. A village meeting was announced, but before it got started, DeAndre talked it over with his wife. She had a solution to the need for more information—the FULRO fighters she had worked with when DeAndre was "captured." They lived near the camp, and some even went into the city. Even better, some of the women had worked inside the U.S. camp as "hootch maids" who cleaned rooms, washed and ironed uniforms, and shined boots.

Relaying that idea to the entire village, it was quickly decided that Chel would go with a small team of fighters to locate DeAndre's "kidnappers" and ask them for more information. But DeAndre was forbidden to accompany his wife. He would stand out too much, imperiling not only himself but the entire team.

The group left first thing in the morning. DeAndre was not used to seeing his wife bare breasted in public, but he also understood she could move much faster through the jungle without vines grabbing cloth.

They returned a short two days later after an uneventful trip. The American departure had been orderly and obviously well planned. Besides the aircraft DeAndre had seen, there had been large truck convoys, taking the Americans to many other places, such as Quy Nhon, Nha Trang, and Saigon.

They had also returned with a real treasure. The hooch maids had privileges at the American PX and three of the ladies had even purchased American civilian radios. After Chel told them of their need for more information, the FULRO village graciously gave one of them to Chel's team. DeAndre was astounded that he could receive the Armed Forces Vietnam Network station broadcasting from the coastal city of Quy Nhon, despite the mountains between the station and the village. Knowing there was no electricity in any DEGA village, the team also brought along a large number of D Cell batteries.

DeAndre didn't waste batteries on music shows but turned on the radio only at the top of the hour for news broadcasts. He also tuned in at 6 P.M. for the hour-long news show.

And the news was shocking. America wasn't just leaving Pleiku—it was negotiating its complete withdrawal from Vietnam at a conference in Paris. DeAndre found out there were no more American combat units in country. As aviation units withdrew, they left the airfields and most of the facilities to the southerners. Hospitals were closing; the 67th Evacuation Hospital at Pleiku had already closed.

While there were Americans still using the huge logistics base at Cam Ranh Bay, they were leaving too as units were able to turn over equipment and facilities to the South Vietnamese. Quy Nhon, where the American radio station was located, was closing immediately. DeAndre guessed once Quy Nhon was closed, the nearest Armed Forces Vietnam Network station would be in Nha Trang—most likely, too far away for DeAndre to hear it. Soon, the only source of news would be Vietnamese stations, but none of the FULRO people spoke the language fluently.

DeAndre had never felt so alone.

As the leaders discussed this and other news, DeAndre realized the Dega weren't as concerned as he was. The hill tribes were self-sufficient and needed no logistical umbilical cord. It was only DeAndre who felt alone.

Just as he was feeling morose, Anh came to visit with some hot intelligence. The Rangers were leaving their camp in Kontum in a hurry. A big hurry, as some were leaving via Chinook helicopter. That meant they couldn't take all their equipment. But the same applied to the Rangers leaving by truck. They simply didn't have space for everything. Anh had no idea why they were leaving, but he also felt it was a permanent move because no guards or the regular ARVN security unit were left behind. They had simply abandoned the camp, leaving a lot of weapons, ammunition, and equipment behind.

The leaders made a hasty plan. They needed to get to the abandoned Ranger camp as quickly as possible, albeit with caution. It was possible—though unlikely—that Thiếu tá Hoang was setting a trap. The fighters were divided into two groups. The first would proceed directly to the camp and begin gathering what they could. The second group would reconnoiter the area to be sure there were no Rangers laying in wait.

Speed was of the essence. They needed to get there before the local civilians stripped the vacated camp bare. It was decided they would travel at night and even risk the road in places where it would be faster than the jungle. The men were given two hours to get some sleep before the mission began.

Sure enough, as the teams approached the old Ranger camp, they saw some civilians scrounging for whatever they could find. But they left the weapons as they had no use for them. The Rangers scooped up all they could carry and departed the area within two hours, hoping Thiếu tá Hoang didn't sic helicopter gunships on them. The trip back to Plei Mrong took a little longer than it took to get there, owing to carrying a lot of weight, and taking a more devious route back in case anybody tracked them.

It was DeAndre who found the best prize. It was another radio, but this one was a shortwave radio, capable of picking up the Voice of America and the BBC. It used the same D batteries the other radio used. DeAndre was sure some of the FULRO fighters saw the radio, but not knowing what it was and seeing it was not a weapon or ammunition, they didn't bother with it.

The warriors were met with an appropriate welcome home. DeAndre faced a choice of either making love to his wife or listening to his new radio. He wisely chose his wife. At dawn, he was ready to turn on the radio, but Chel had other ideas. After two other times during the night, he thought to himself, "Oh no, not again" as she mounted him. But soon he forgot the radio in a spasm of erotic sensations. He was still forgetting the radio as the two lay together snuggling in afterglow.

As Chel and the other women prepared breakfast, DeAndre fired up the shortwave radio and soon found the Voice of America as it broadcast news about Asia, and in this case, mostly about Vietnam. The news from Paris was that the negotiations were going well but there were no specifics. America's withdrawal continued, with the stated goal to have all Americans, military and civilian, in Saigon. DeAndre assumed Smithy and Lee Anne were there, finishing the translation of the New Testament into Jarai.

It was obvious from the news that the Paris Peace Accords were intended to stop the fighting and established the International Commission of Control and Supervision. The four nations selected to serve were Hungary and Poland, representing the Communists, and Canada and Indonesia which looked after the interests of the non-Communists. It was hoped that the treaty would be signed sometime during the next month, January 1973.

The Americans may have been planning to go home, but the Degas were already home and could not escape the fighting. Four days after the raid on the old Ranger camp, a listening post stationed about a mile from Plei Mrong ran into the village. He announced that he had spotted an ARVN force coming from the direction of the road. Most likely, they were on the way to raid Plei Mrong. The lookout estimated it to be a company-sized unit of about 100 men.

Due to DeAndre's leadership, the FULRO had made contingency plans for just this kind of event. There was no time to get the women and children out of the village, but instead, they were hidden in well-camouflaged bunkers in the jungle near the village. The men took their assigned positions. There had never been any thought given that DeAndre would not be among the fighters and he took his own position near Mip. They were all very well concealed. The plan was to allow the ARVN into the village, then open fire with Mip firing the first shot.

All the FULRO on the east side were inside of structures in places that offered both concealment and cover. The men on the other side took their positions in trees that ringed the village. This allowed the fighters on the ground to shoot straight into the enemy without worrying about hitting their teammates. Some of the tree shooters also deployed claymore mines near the base of the tree they were in. A small detachment was dispatched to the road. They would take out any ARVN stragglers who escaped the carnage in the village, as well as the truck drivers.

DeAndre watched the vanguard of the ARVN force approach the village. They didn't know if the tribesmen had fled or were still in the village. They halted and scoured the hootches. Soon, an officer appeared, radioman at his side. It was his decision to make, and the men cautiously moved forward into the village and were met with dead silence. After a few minutes the demeanor of the ARVN changed from super cautious to being very relaxed. They thought all they had to do was plunder the empty village and leave when they wished. Even the officer seemed to exude an attitude that this would be a cakewalk.

Since rank has its privileges, he selected the building he wanted to search and began to move towards it. Of course, it was the long house, but it was also where Mip and DeAndre lay hidden. With nods of their heads, they decided Mip would take out the officer and DeAndre would make sure nothing was radioed back to headquarters.

The ARVN Trung uy put his foot on the first step into the long house when Mip opened fire with a shot to the chest. A split second later, the radioman also pitched forward, a bullet in his heart. Both Mip and DeAndre made sure their targets were totally down before looking for other targets.

All hell broke loose on the ARVN soldiers standing leisurely in the center of the village. Most did not return fire either because they could not identify where the shots came from or they were dead. Some few recognized where the fire was coming from and retreated in the opposite direction—right into the fields of fire of the tree snipers. Some few of them managed to get to the base of a tree, only to be blown apart by a Claymore mine.

But some ARVN did manage to return fire. It wasn't very effective, and no Degas were killed, but three were wounded. One of the wounded was DeAndre.

A few other ARVN, including some of whom were lightly wounded, ran towards the road and their trucks. The tribesmen let them go. Scared and literally breathless, the ARVN ran towards the trucks, yelling at the drivers to start the engines. Too late. The FULRO team had designated some of their group to take out the drivers. As the tribesmen opened fire on the escapees, the drivers slumped forward while sitting at the wheel of their trucks. There would be no word of the ambush getting back to the ARVN base.

Then began the process DeAndre disliked so much, even though he understood why the Dega did it. As he watched, FULRO fighters administered the coup de grace to the wounded ARVN soldiers they found. The slaughter was complete.

Only after the ARVN had fled the village did Mip see that DeAndre was wounded and in pain. One of the village men who had been trained as medics by the American Special Forces, came to DeAndre. His halting English was good enough to convey to DeAndre that one round had gone through the fleshy part of the right thigh and did no real damage. However, the right knee has been grazed. While the medic knew the kneecap was intact, he wasn't sure what else had happened. He suspected there was a small bone chip from the kneecap that was now causing pain in a tendon.

About that time, Chel appeared. DeAndre could not have been prouder of her. Rather than wailing and getting in the way of the medic, she talked the situation over calmly with both her husband and the medic. All three agreed the medic should make a small incision and try to locate the chip and remove it—provided, of course, that there really was a bone chip.

That incision would be painful because there were

no painkiller drugs in the village—not even aspirin. The medic didn't want to operate without some sort of dulling of the pain. Doing so would risk DeAndre twitching and moving at just the wrong time, causing more damage.

It was Chel who had the idea. She left DeAndre's side, then brought Ama Tum back with her. Grinning in a celebratory way, Chel said, "We know how make pain gone."

With that, Ama Tum revealed a jar of rice beer. Knowing the Jarai Christians eschew alcohol, DeAndre looked puzzled. "Use for Communion," said Chel, "but Ama Tum say should so you no hurt."

DeAndre grinned. "I had to get shot before I could get a legal drink," he chuckled.

"What 'legal' mean," she asked.

The knee patched up and bandaged in no time, Ama Tum, Mip, and DeAndre got their heads together. Though it would take some time, the ARVN would sooner or later realize an ambush had wiped out one of their units and they would eventually find the village. The ARVN would be vengeful, the FULRO leaders knew they must move Plei Mrong very soon.

The three leaders quickly came to a plan. Three small teams of three men each would go deep into the jungle, each team tasked with finding new locations. Once that was done, they would all return to the village where a choice would be made.

Ama Tum added that rather than dictating how long the teams would stay in the jungle, they were rather instructed to return "as soon as possible" after finding two places. While they were gone, the remainder of the fighters, the old men, the women—and DeAndre—would prepare to move as quickly as possible.

For once, DeAndre knew when to quit. Both because of the pain and the bandage keeping him from flexing the knee, he could only hobble around. He most certainly could not function as a fighter in the jungle.

Because the Dega had always led a semi-nomadic life, they had no televisions, kitchen stoves, family China, or other possessions that were difficult to move. Even their homes could be rebuilt quickly, while such household items as cooking utensils, foodstuffs, machetes, and sleeping pallets could be moved easily and quickly, though everyone—even the children—would have to help carry things. During each of the three nights the teams were gone, they gathered for prayer with only the dim light of their primitive lamps. Some men did not attend, being needed to provide early warning should ARVN troops be spotted.

The teams returned over the course of the third day. They had been instructed to go to different areas of the jungle—one nearer to Cambodia, the second further south, closer to Rhade villages, and the third closer to Kon Tum.

The entire village, including the women, participated in the discussion. They all listened to the teams, with many villagers recognizing the places recommended and adding either their endorsement or opposition to the suggestions. Slowly, the peoples' preference was for those new locations nearer the Cambodian border. As Mip pointed out, once the north won the war against the southern Kinh, their only enemy would be the Montagnards. Being closer to Cambodia would give them an escape route should it be needed. Besides, there were some Montagnard villages across the border.

But before a decision could be made, prayer was necessary. That evening's prayer service was devoted to asking God's guidance. After the service, Ama Tum asked everyone to pray as a family, and the village would decide the next morning.

The decision was surprisingly easy. Most felt that one of the locations near the border and was decently far from roadways was the best. No actual vote was taken. All accepted the decision of the leadership to move to that place and set up the village again.

Messengers were sent out to other FULRO camps telling them of the new location. One messenger went well to the south into Rhade country to tell the FULRO leaders of their new location, as well as learning about other fighters.

He returned to Plei Mrong with a message to its leaders: send them to Lo Mo for a FULRO-wide meeting of head men. The discussion would be about plans for FULRO now that the south's fall was imminent. The gathering was slated to take place during the last week of the month, so Ama Tum, Mip and DeAndre would leave once the village had moved.

Preparations were finalized that day. The village was to be moved early the following morning. Loads were distributed, the young and strong helping the older and weaker. DeAndre didn't like this part, but it was true he could not carry as much as he normally would, so he acceded to letting Chel get a few of the smaller children to carry their household goods. That night's prayer service was spent asking for two things—safety for the move and guidance in giving the village a new name.

Two days later, DeAndre and Chel walked into the small clearing that would become home. They had already crossed the nearby Prek Sathay, which was still a small creek. While the site was less than six miles from Cambodia, it was also fairly near Route 14C and about equidistance to Kon Tum and Pleiku. This was the new home.

Chel busied herself setting up a small campsite that would be home until the new longhouse was built, while DeAndre contributed his part by making a ring of stones for the cooking fire. Like the other women, she walked down to the creek to get some water to make a simple evening meal of rice cooked with some beans and peppers. Before sleep, everyone except some guards met under lamplight for prayer, thanking God for the safe trip and asking His guidance in finding a name for the new hamlet.

DeAndre was awakened in the predawn hour by his wife's kiss. She knew, as did he, that time was of the essence in getting the village up and running. While she busied herself working with the women to make a prayer place, DeAndre joined the men to stake out the longhouse. By first light, progress had been made on both counts.

Breakfast was followed by prayer. Ama Tum asked them to stay in place after worship. DeAndre proposed they keep the name Plei Mrong. Noisy chatter alluded to a quick acceptance of the idea.

By midday, Ama Tum pulled DeAndre away from his work at the longhouse. He wanted the three village leaders to discuss when they would find a new battle camp location. It was decided that since DeAndre's knee was much better, he would lead a group of four fighters the next morning. The Jarai would scout out locations and DeAndre would select the best one.

Chel objected to her husband being in the bush on his bum knee. DeAndre knew she was just being protective, so he had to be gentle in telling her he was going anyway. Chel had no choice, but after pouting a little, she agreed. She wished they had privacy, but they still slept in the open, as did the rest of the village. Besides an extra-long passionate kiss, she could not properly send her beloved on his way.

The group left in a light rain the next morning. That was a good thing as the rain would mask their movement as well as ensure any Kinh soldiers were bundled up against the wet.

While the new Plei Mrong was close to the border, the group felt they needed a battle camp further away in the other direction, keeping them closer to the northern army. The camp also needed to be a happy balance between being close enough to the road for ambushes, but not so close that the enemy could sneak up on them. After finding the first possible location, they left DeAndre there while the others singly went off to find more places.

As they returned, they shared what they had seen, then agreed to visit each on the following day.

DeAndre liked the second place best. It had multiple escape routes, the fighters could post lookouts along the road, and it was the most defensible. They returned home on the third day, telling Ama Tum and Mip of their recommendation.

Those three days had been well used. The longhouse was taking shape; some pig sties had been built; and a small garden cleared. Squash, eggplant, and peppers had already been planted. The prayer space was ready and being used, a new cross having been fashioned and erected.

Those families who would not live in the longhouse had started erecting wooden frames and begun putting wattle up for walls and broad palm leaves for thatched roofing. Ama Tum decided the men would work on the village for another three days before setting up the battle camp and resuming military operations.

Only then would Ama Tum, DeAndre and Mip leave for LoMo for the top-level meeting of FULRO. In the meantime, lookouts were concealed along any likely routes the Kinh soldiers might take from the road. Soon, the long house was far enough along where the interior could be set up. DeAndre joined other men building walls and doors, granaries, porches, and bed frames. It was beginning to look like home.

But of course, making it homey depending on the women. Lamps were fashioned from tree knots and pine tar. Wall hangings went up and clothing storage places created. The men finished up outside while the women finished inside. The three days were over, though more work was needed. Ample praise and thanks to God flowed upward that evening.

By himself later that night, DeAndre pondered the situation. They were now building the fourth "Plei Mrong." Yes, the village had achieved some excellent military sorties, yet it was also true that each move had been deeper into the jungle and closer to Cambodia. He chuckled to himself. He knew FULRO was in the same position as the Americans during the war against the Communists. True, the Yanks never lost a battle, but they lost the war. FULRO kept inflicting serious damage on the Kinh, yet they also were slowly being weighed down by the overwhelming force arrayed against them.

DeAndre finally unpacked his broadcast radio and was surprised he could get the AFVN repeater station in Nha Trang. He reckoned it must be because of the higher elevation of the new Plei Mrong.

What he learned alarmed him. It was now February 1975. The Americans were all but gone, leaving only the Marine guards at the embassy in Saigon, the Armed Forces Radio staff, and some few assigned as pilots for the Huey helicopters used by the International Commission of Control and Supervision. The southern army seemed to have gone into a fetal position, withdrawing to the big cities.

Other news drifted in. A messenger had left Plei Mrong and gone to Pleiku to inform Anh of the village's new location. The Rangers and other ARVN units that had been in Pleiku had been moved to the coast to help in the defense of Quy Nhon and Nha Trang. Many civilians had moved with them, including their spy, Thuy. Lee Anne and Smithy had left some time ago and were now in Saigon.

Units of the People's Army of the north were moving onto the military base and into the city. Anh said he and his family would be leaving soon. As members of the ruling class in the south, Anh knew they would be arrested and persecuted. He knew he was especially in danger because he had lived in the United States. Anh's father had been in contact with some southerners who carried escapees to the Philippines. The group operated out of small coastal villages between Quy Nhon and Tuy Hoa.

The village triumvirate knew their only enemy was now the northern Kinh. Once Saigon fell, the Communists would look to the mountains as the last place to conquer. They wanted the Central Highlands as much as the southerners had and for the same reasons—to be able to grow coffee and tea for export and because they thought of the mountain tribes as moi.

DeAndre risked running out of batteries for his short wave radio, but he needed information about the fall of the south. Keeping it tuned to the Voice of America, he understood plans were being made to evacuate Saigon before Communist troops entered the city. He heard of one last major battle between the ARVN and the northerners at Xuan Loc. When it fell, it was a foregone conclusion the south was finished.

It was near the end of April when DeAndre heard a very strange song come from his radio: Bing Crosby's

"White Christmas." He wondered if it was some sort of code for the Americans.

Sure enough, the next day, the Voice of America announced Saigon had fallen. The Communists had won. A few days later, two of the FULRO fighters came back to Plei Mrong after spying on the Communist troops at Pleiku. The victory celebrations had been over the top with drunkenness, shooting into the air, and eating food stolen from civilian homes.

CHAPTER NINETEEN

DeAndre knew the Dega would have a short respite before the armed might of the People's Army descended on them. The number of lookouts on the road was increased.

But, what to do? Wait until the Kinh attacked them? Be aggressive and attack them first? Attack the "New Economic Zone" villages? It was decided that before the Plei Mrong fighters did anything, they would wait until Ama Tum, Mip, and DeAndre returned from the FULRO summit. An early morning prayer service took place to bless the village leadership going to the FULRO meeting. The villagers asked God for the trio's protection as well as for wisdom on the part of all those who gathered in Lo Mo.

Chel woke her husband up a little earlier than necessary for him to depart on time and gave him her own special send off. She was not worried about him as though he was going to battle, but she knew she would miss him greatly.

The three stayed off the roads to avoid being spotted by Communist aircraft. Instead, they used the network of footpaths used by the tribes for years. Each man carried enough food for the requisite three days. DeAndre had been with the Jarai long enough to allow him to pick up some skill at navigating the thick forest. Mip assumed the lead as he had been to Lo Mo before.

DeAndre had come to enjoy the jungle. The thick canopy of tree leaves above them kept them cooler than if they'd be walking on the road. He was learning what the various jungle critters looked like and found he was losing his fear of snakes. They did indeed want to get away from him as much as he wanted to avoid them.

Though he was with two other men, they seldom talked, allowing DeAndre to commune with nature. On the second day, he had a bit of a revelation. His new connection with the creation began to turn into a connection with God. On day three, as the steady pace of their walk turned into some sort of mantra, he indeed had a link with God. He quietly savored the new relationship, realizing walking was now his prayer time. Religion was taking on a more personal tone.

It was late afternoon when they arrived at Lo Mo which was deep in the jungle, as were all other FULRO villages. It was a little larger than most, simply because it functioned as the headquarters of the rebellion. A guard had stopped them before they entered, and they waited quietly until FULRO leaders identified the three.

They were led to a longhouse where they ate the evening meal with other leaders, giving them all a chance to get to know each other at a more personal level. Many seemed puzzled by DeAndre's presence. It wasn't so much that they had never seen a black man, because most had been with Special Forces teams that included black soldiers. In his halting Rhade, DeAndre explained that he was an experienced combat officer who had married a Jarai woman and was dedicated to fighting the Kinh. The pleasant evening ended when they were given sleeping mats for the night.

DeAndre's eyes were glued shut by the normal dried eye boogers when he was awakened by a gentle hand shaking his shoulders. One of Lo Mo village men awakened the three before dawn, telling them that breakfast was almost ready.

Breakfast included almost everyone in the village, preceded by prayer spoken in Rhade. DeAndre was surprised he knew what was said.

"Good morning" said a Rhade man in passable Jarai. Standing in front of them, he continued. "My name is Y Hinnie and I welcome you to the FULRO leadership meeting." Ama Tum thanked him for using Jarai but went on to say that the three had been learning Rhade.

Y Hinnie led them to a semi-secluded place under a large tree where a small group of men had gathered. In their halting Rhade, DeAndre, Ama Tum, and Mip introduced themselves and were met with smiles and welcome.

The meeting started with prayer followed by an intelligence brief. DeAndre was surprised at how much FULRO knew about Kinh forces. Trucks full of soldiers were already moving from the coast into the mountains. It was also reported that the Communist government was planning New Economic Zones, a thin disguise for creating new villages in the mountains for South Vietnamese who had been displaced by victorious northerners who had confiscated their property. It wasn't known yet if troops would protect the new villages or just remain in their bases, nor were locations known for any of the New Economic Zones.

What followed was essentially a brainstorming session. What would FULRO do? Attack the troops? Attack the villages? Set up ambushes on the roads? Destroy any crops the new villagers planted? There was a lot of discussion, but no discord. The conversation continued into the next day, breaking only for food and sleep.

Finally, it was decided to attack the villages as the primary tactic. If the target was too big for the nearest FULRO village to handle by itself, coordination would be made with other teams.

Though the three from Plei Mrong, as well as the other representatives from other villages, could have left to go home, Y Hinnie invited everyone to spend the evening meal with the Rhade and stay the night.

While walking back to Plei Mrong, DeAndre heard truck traffic on a nearby road. "Let's go see what they're doing," he suggested to the others. As usual, they stayed well concealed as they approached the road. The trucks had disappeared up the road, but they heard another convoy approaching. There must have been twenty trucks, each filled with civilian Kinh.

Whole families, including small children were aboard. Once the convoy was out of sight, the three began tracking the convoy in an effort to find out where the new village was located. They tracked for a few hours but were interrupted by yet another convoy. Slipping into concealment, the three counted another twenty trucks, and followed that convoy as well.

Such were the men's skills, that they were able to track the trucks even after nightfall. Thinking they had no enemies anymore, the convoy leaders had the trucks disgorge their human cargo, then begin the return trip even after dark.

The trail ended a day and a half after tracking had begun. The Vietnamese civilians had little in the way of construction material. Most of them had been city dwellers and had no idea how to make a thatched roof, much less construct walls from bamboo.

DeAndre seemed to think the site was near the Cambodian border. After a short reconnaissance, the others agreed. Noting the new Kinh village's location near the road, they then continued their trip home.

Their homecoming was everything they had hoped it would be. The construction of the village was finished and some farmland started. They could not use "slash and burn" simply because the smoke would have given them away. But there were a few small clear areas, and they were converted into fields, a different crop for each. Care was taken to plant the fields irregularly so it didn't appear to be a nice tidy farm to observers in the sky.

After breakfast, Mip convened a meeting of all the fighters. They shared the news from the meeting as well as the finding of the new Vietnamese village. The first thing they decided was to send a messenger to Lo Mo informing the FULRO leaders of the new Vietnamese settlement.

But—what to do. “We could ambush the convoys on the road,” said DeAndre. Mip continued, “Or we could attack the village, with the most important thing being to destroy any infrastructure. I don’t like the idea of killing civilians.”

“I wonder if the government will build up defensive systems within the village and train the men as soldiers.”

“Or, maybe they will set up small army posts around the village to help protect it.”

“I wonder if they know we are still here. They might just leave the villages unprotected—at least until we hit a few.”

“Let’s send a recon team to see if anybody is armed.”

The conversation continued almost the entire day, with the agreement being that they would find out more about the Kinh village first, but the overall strategy would be to attack the villages rather than tangle with troops.

A team of four left the next morning to spy on the Kinh, mostly to find out if they were armed, and if so, was there some sort of organized quasi-military group that had been trained to defend the village.

Chel wanted to know if DeAndre would be a fighter again. If he was seen by enemy troops during a battle surely he would stand out enough to be reported up the chain of command. Once his presence was noted, the enemy would know he was an American and try their damnedest to find and kill him.

“Yes,” he replied quietly. “I will fight. I cannot stand by and let the other men fight, especially since I have a lot of experience in combat.” Though frightened, she knew the other wives were worried about their men too and accepted the situation.

DeAndre looked up from the squash plants he was tending and saw the four scouts return and go into the long house. Telling the other men he had to leave, he headed to the meeting taking place at Mip's.

The Kinh village was pathetic. With neither building materials or knowledge, city folk had been able to build only lean-tos for shelter. They seemed to have plenty of food which had come along with them and they had adequate clothing. But, after three days of observation, and seeing no weapons, the team was about to depart when a single truck appeared, loaded with ten men and some simple tools.

The new arrivals had the weathered look of farmers, not the light-colored skin of the city dwellers. The spies stayed for two more days and saw the ten men instructing the settlers how to build simple huts of wattle walls with thatched roofs. By the time the team left, several huts had been built and the first rudiments of a garden had begun.

The leadership team heard all this with great interest. After thanking the spy team, the three leaders decided to develop a suggested course of action, then call a meeting of all the fighters and see if they accepted their plans.

To begin, Ama Tum told the villagers what was known about the Kinh hamlet. There were no guns, and the instructional team had no weapons either. They were all city folk with soft hands. They'd brought their children, which indicated a permanent move. There didn't seem to be any troops in the area.

DeAndre had been surprised how well and how quickly the Jarai had taken to the idea that everyone was welcome to give input into plans. Nobody was keen on killing unarmed civilians, yet at the same time, they needed to destroy what little infrastructure had been built. They thought it wise to wait a couple of weeks to allow more infrastructure to grow as well as give the Kinh time to let their guard down.

The time had come, and the FULRO stayed concealed as they watched the Vietnamese village. Indeed, many more huts had been built. Before assaulting the village, the men all lit their little pine tar lamps, even though it was daylight. On a signal from Ama Tum, the Jarai fighters moved quietly into the open center of the village, weapons at the ready. Women screamed and children buried their faces in their mother's clothes as the men looked on feeling helpless, but defiant nonetheless. DeAndre remembered the look on the faces of the Jarai men when the Rangers used to raid their villages.

The Dega checked all the huts to be sure no fighters were concealed, then began to use their little lumps of flame. The wattle huts burned quickly, the crackle of flames mixing with the sobbing of the women who watched what little they had in the world burned to the ground.

As soon as they saw all the huts were burning, DeAndre and the others left, blending back into the surrounding jungle. The garden was untouched.

DeAndre wondered how long it would be until communist troops arrived. Most likely, this would be the last easy attack they would get. It would take messengers from the village some time to get back to Pleiku to inform the military.

The walk back to Plei Mrong was easy. It's always easier to move when there is good news that lightens the heart, and the ease by which they burned the Kinh village was good news. DeAndre was proud to be a part of such a group, fighting for their freedom, even though he was a different race, culture, and country. He had to smile at the men's clothing, some in traditional Jarai loin cloths, others wearing decidedly western clothes, and some mixing the two. The man directly in front of DeAndre was a good example. He wore a New York Yankees baseball cap (badly faded) plus an old camouflage uniform shirt covering a Mickey Mouse t-shirt.

Few of the men slung their rifles, simply because the slings had disappeared some time ago. Instead, they carried the weapon in one hand with the muzzle pointed forward and down. Most had a bandolier of ammunition across their chest, while others carried their extra ammunition inside old Claymore mine bags. A motley but dedicated crew.

Their welcome home was all it was expected to be… joy in the men's safe return and joy in a successful mission. But Chel had more joy to share with DeAndre. After the more public welcome was over, she herded him into their room and closed the door.

After a long and passionate kiss, she had him sit on the bed while she stood in front of him. "I have news," she said. "I hope you like." His puzzled look invited her to continue. "I have baby."

DeAndre didn't think a human could experience so many emotions all at once, but the most prominent response was boundless happiness. After a moment of incredulity, he stood up, hugging his wife and saying, "I love you" over and over again. He was going to be a father! He wished so much he could share this moment with his mother.

The two went back outside and joined the rest of the village on the way to the evening prayer service. He had trouble keeping the news inside him until the right time. Finally, the part of the service came—the part where people told what they were thankful for. Most stood up to express gratitude for a successful mission, plus thankfulness for everyone's safe return. Then DeAndre stood up. Pulling his wife to her feet next to him, their faces shining, Chel announced her pregnancy. The shouting was so great that few heard her thank God for the new miracle.

Before drifting off to sleep that night, he talked to God. On the one hand, he most certainly didn't deserve the many blessings he'd been given. On the other, he was happy to worship a God who didn't give him what he deserved.

The village's lightheartedness lasted two weeks before lookouts reported heavy troop-carrying truck traffic on the road. The war was soon to begin in earnest as the Vietnamese would try to wreak revenge for the burning of their fledgling New Economic Zone village.

It came without warning. As the fighters departed

the battle camp, it was consumed in a huge fireball that spread out and engulfed a few of the Dega men. Hearing the jet engine power the aircraft in a fierce climb, DeAndre knew napalm had been dropped on them. But the rest of the men had no idea. Not all the men who had been hit were dead. They were running aimlessly, screaming in pain. DeAndre chased after the closest, tackled him, and quickly rolled the victim over and over in the dirt. He was alive, but when DeAndre saw the extent of the burns, he almost wished he hadn't saved the man. He would not live much longer, but that time would be spent in unimaginable pain.

Making a snap decision, he did nothing to try to help the man, except to speak gently and softly in an attempt to make things a little easier for him. DeAndre looked around and saw he had been mimicked and other napalm victims were on the ground, surrounded by their friends. After a few moments, his man lapsed into consciousness, making him oblivious to his badly burned body.

DeAndre got up and walked to other groups and saw that some of the victims had already died. The uninjured asked DeAndre if there was anything that could be done for the victims, but he merely shook his head. The original group at the battle camp had numbered twelve, but now five of them lay dead or dying. This was the first multi-casualty fight the Plei Mrong fighters had suffered, and they hadn't even had the opportunity to fight back.

But what to do next. "The bodies need to go back to their families," thought DeAndre. Lik (say leek) saw DeAndre off by himself. He was the other patrol leader with

DeAndre. Speaking Jarai slowly so that DeAndre understood, he said, "Yes, dead need go back to Plei Mrong, but trip be very slow, very noisy, and bad security. Maybe bury here." The two talked for a few more minutes. DeAndre totally understood Lik and agreed with him.

Lik being Jarai himself, he would better know the reaction of the people in the village. The decision was made—the men would be buried in the jungle with heavy triple canopy to shield the graves and the worshippers. None of the men even hinted at any disagreement with the decision.

Lik dispatched two men to find a suitable place. They returned within the hour, saying a good burial site was close by. Litters were fashioned from ponchos, and the men began the sad task of taking their buddies to their final resting place.

Digging the graves was hard work. Their short-handled military entrenching tools were a poor substitute for shovels. As daylight began to fade, the work was finished. DeAndre began with a prayer of thanks for the lives of the men who died, then turned it over to Lik,who followed with a prayer of pleading, asking God to take their friends. One of the other men ended the very short service with a few words, asking God to let all of them be reunited in heaven. With that, the men solemnly shoveled dirt into the graves.

Normally, the Jarai didn't travel through the jungle at night, primarily being nervous about tigers. The only way one could know if there was a tiger nearby was when big yellow eyes were seen peering through the bush, but by then it was too late. But they overcame those fears so they could return home and let wives know of their husband's death. DeAndre also wanted to talk to Ama Tum and Mip soon. There was a lot to decide.

Chel was just beginning to show a baby bump, but with DeAndre's permission, she went with the other wives to console the new widows. DeAndre listened to the wailing, heard other children asking questions, and knew that a new and tough time had arrived for FULRO.

He found Ama Tum and Mip and the three talked for hours. "I don't think there is any reason to downplay this," said DeAndre, "because these are strong people, imbued with a desire to fight for the freedom of the mountain people and for their own children."

He quickly found he didn't need such rah rah talk. They agreed quickly, saying only that whatever was decided, they needed to involve the entire village as quickly as possible.

The decision was long in coming. There were a lot of options, most of which included the other FULRO villages. They asked one man who had just returned from Pleiku to come by and let them know what he'd seen. The Kinh had come down hard on the minority tribes and one of the main things they'd done is burn down all their churches. The ownership of Bibles was forbidden and that was enforced by police coming into homes without warning and looking for the holy book.

The Jarai population of Pleiku was shrinking as more people realized that the Communist strictures would just get worse. Some decided to stay, but the Jarai also knew those were dedicated FULRO members who wanted to remain and spy on the government.

The sentinels on the main road had also seen an increase in the number of trucks bringing more civilians into the mountains, undoubtedly to build new Kinh villages.

There wasn't time to travel all the way to Lo Mo to consult with the main FULRO leadership, but even though darkness had fallen, two runners were sent out to those FULRO villages closest to Plei Mrong.

Should they continue having a main village and a battle camp? Should the families be moved into Cambodia? Should they do low risk harassment raids or full-blown attacks on the new Kinh villages? How would they be resupplied with ammunition? They were still deep into conversation when Chel and the other wives interrupted to tell the men that it was almost dawn.

A few hours' sleep, then a morning prayer meeting with the entire village.

There was certainly no shortage of ideas. The discussion lasted until the noon hour when DeAndre said the time had come to decide—no more new ideas. Ama Tum agreed.

After a short break for food, DeAndre now recited the main ideas that had been offered by village members. As to ammunition, it was decided to send two fighters to Cambodia to contact the Khmer Rouge and ask for help. After some sharp debate, it was also decided there would not be a separate battle camp. As to strategy, the decision was to start with simple harassment raids that would minimize casualties and conserve ammunition. If there were to be bigger attacks, they would be done in conjunction with other FULRO fighters.

Should they move Plei Mrong? Had the Kinh spotted it? Would it suffer the same fate as the battle camp?

Plei Mrong would indeed move—and keep moving. There would be no more longhouses, prayer rooms, or other permanent buildings. The villagers would build only simple lean-tos, including the prayer house. Gardens would not be located near any new settlement, but rather there would be three gardens that would be attended to regularly. There would be no more livestock except chickens. The pigs would be slaughtered soon and the few that remained would go to the first new location.

In short, the people of Plei Mrong would become nomads, moving frequently. It would be a hard life.

Though DeAndre agreed with the decision, he also knew his wife was pregnant and the moving would be all the harder to do as her pregnancy advanced. She knew her husband was concerned, but she also told him she was tough. Tougher than American women. That made him laugh, though it did little to decrease his worry.

Runners went out to Lo Mo to let the FULRO leadership know of the constant moving. Another small team went out to find the location of the new, skinnier Plei Mrong. A third small team went out to scout other New Economic Zone villages to raid. The women organized themselves to support the five new widows and their children. Everyone understood that some time-honored gender roles would have to change. The men would be leaving the village more often to fight, and all understood that some would die in battle. The women would have to fill in as best they could, in addition to doing their own work, all the while nomadically moving.

One of those gender roles was fighting. With the men out of the village more often, there was the risk of Communist soldiers hitting Plei Mrong while they were away. The women needed to learn to fight, though they would not go out on missions. Their sole purpose was to defend the village. Some old Chinese-made SKS rifles were brought out of hiding with the intent of teaching the women to shoot.

The SKS had the advantage of using the same ammunition as the Kinh's AK-47 and was a bit lighter. DeAndre selected three other men to help him teach the ladies. While the class was learning how to break down, clean, and maintain the weapon, a small scouting force was deployed to be sure there weren't any Communists nearby that might hear the shooting.

DeAndre realized the women would only be shooting short distances because they would be in the jungle village. They could forgo teaching the women to shoot at distances beyond 100 meters. The men fashioned some humanoid targets from palm leaves. After five half-days of shooting, DeAndre pronounced them ready.

Henceforth, anytime the men were gone—or even if only half of them were gone—the women knew they were mothers, cooks, launderers—and fighters. Each time the village moved to a new location, a shelter was built for the children. In the event of Kinh soldiers raiding the village, some women would care for the children and act as last chance protectors, while the others actively fought the enemy.

The two men who had been dispatched to Cambodia in hopes of obtaining ammunition from the Khmer Rouge returned empty handed. They had found only one small Khmer village, but there were no fighters. It had been difficult to communicate with the villagers, but they made it clear they did not know the location of Khmer Rouge warriors

As Chel's time grew closer, the more concerned DeAndre became. A midwife had been designated to help with the birth. Though he wanted to be there to help his wife, she and the midwife were firm in their decision that he not be allowed to be around Chel until the baby was born, cleaned up, and in its mother's arms. That "no men allowed" policy had always been the norm. When DeAndre bemoaned it, the other men just grinned and said that was the way it was supposed to be.

When a scout team returned with information about a new Kinh village being built a half day away, the leadership decided to raid it. The scouts had not seen any troops in the vicinity, but they did see that some of the men were armed. DeAndre wanted to lead the raid, but Ama Tum and Mip said no—he needed to remain in Plei Mrong to support his wife. He saw the wisdom and gave them only a token argument.

He joined the other villagers and Chel as they met in the prayer room (now just a place where a cross was planted) and asked God's blessing on the raiding party. As the men left, he watched with pride as the women calmly and professionally slung their weapons and rehearsed defensive strategies, all the while going about their daily activities.

Lik saw the file of Jarai men ahead of and behind him, each carrying an AK-47. A member of the scouting team that had found the Kinh village led the way, halting the column well away to negate being spotted. Lik and three others reconnoitered the settlement.

Sure enough, some villagers were armed, but there was no evidence of fighting positions, though Lik spotted some places that would provide decent protection. Like similar villages, the people had not built permanent housing yet but still lived in lean-tos and similar temporary shelters. The team also found the men planting coffee plants in what appeared to be a clearing made by Montagnard's old slash and burn methods.

A strategy was quickly developed. A very small detachment would open fire on the few men in the village, while the main force waited for the farmers to respond. The Kinh would have to pass through a narrow forest path and that would make an excellent ambush site. It was planned that the attack would take no longer than five minutes, at which time the Dega fighters would disengage and move away quickly in a direction away from Plei Mrong.

The plan worked to perfection. The first shots into the village downed some Kinh men. The remaining men were not trained soldiers and reacted accordingly, standing up and looking around rather than seeking cover. The FUL-

RO fighters downed a few of those too, all the while the women sought shelter in the lean-tos. The shooting lasted less than a minute.

The farmers obviously were not trained soldiers. With no thought to good movement security, they ran pell-mell towards the village through the narrow passage where they were met with a hail of gunfire. It was a good minute before the first of them returned fire, but since they didn't know where the Dega were hiding, it was scattered and ineffective. Lik saw that the number of targets had declined and decided to end the ambush after only four minutes.

Not a single Jarai had been hit. The small detachment of FULRO fighters from the village joined the main group and everyone made an orderly and quick departure from the area, moving away from Plei Mrong.

The patrol meandered for two days to be sure they were not followed, then moved towards home. Chel led the welcome committee of happy wives.

But the fighters had a growing problem—how to get more ammunition. They hadn't been able to get any since the Rangers had pulled out of Pleiku almost a year ago. Remembering that the villagers had decided to build the new Plei Mrong near the Cambodian border, the leadership decided that one of them needed to cross the border in search of a way to establish a regular logistics system for ammunition.

Ama Tum and DeAndre discussed who would go. "Ama Tum," started DeAndre, "you're the only person in the village who has ever talked to the Khmer Rouge. Not only that, but you speak languages other than Jarai. I suggest you go." There was only a little discussion, but Ama Tum saw the wisdom of DeAndre's argument and took on the mission. Taking just one other fighter with him, he departed the next morning.

It was good that he did. In the afternoon, an airplane began circling overhead. Obviously, it was some sort of Kinh spy craft and because it had circled, most likely he had spotted the village. It didn't take but a few moments for Mip and DeAndre to decide to evacuate the village as quickly as possible. One of the preselected sites was designated and within thirty minutes, the entire village was on the move.

Thirty minutes later, the village behind them erupted in explosions and fire. They heard a jet as it climbed away from the scene. Not hearing any more explosions or aircraft, DeAndre knew the bombing was over. He decided to send one of the younger fighters back to watch the old village site for any sign of further enemy activity.

It was very late in the afternoon when the people reached the new site, which was a short distance inside Cambodia. DeAndre dispatched a group of fighters to set up security around the camp, staying concealed in the jungle. The remaining men began preparing simple nighttime sleeping areas for each family, while the women prepared food. As the last mouthful was eaten, all light vanished and the jungle was quiet around them.

Daylight allowed DeAndre to see that the jungle canopy was thick overhead. He didn't know if the Vietnamese would bomb villages inside Cambodia, so it was essential that seeing the village from the air would be well-nigh impossible. Most of the people worked hard at building some sort of inhabitable village, except a few men acting as security, concentrating on the most likely enemy advance route from the nearest road.

In the afternoon, the fighter who had stayed behind at the old village site reported to DeAndre. No enemy troops had been spotted or heard. Remembering that Ama Tum would have no way to find the new village, he'd left a note in Rhade saying only that they were now in "Location 6." DeAndre was grateful the young man had remembered Ama Tum and told him so.

By afternoon, the prayer area was ready, replete with a simple cross made of bamboo. God was thanked and praised that they had escaped harm and asked His blessings to continue.

Mip indicated to DeAndre that the villagers would need at least a week before military operations resumed. That was fine as he watched his wife, obviously about ready to pop. As a first-time father, he was worried and told the midwife that. The older lady smiled at him, assured him that the baby and Chel were both fine, though it wouldn't be long before she delivered.

In the meantime, he brainstormed with Mip and Lik on any future offensive operations. The first decision was to send a messenger to Lo Mo, letting the FULRO leadership know of their change in location as well as to pick up the latest intelligence.

DeAndre wondered if the Communists would continue to leave the new villages without direct military protection.

One of the village women approached DeAndre and smiled. "You are a father." He followed her to the shelter where Chel lay resting with a brand-new baby boy in her arms.

Brao was with them, a product of their love. Both Chel and DeAndre were determined to be good parents—and provide a chance for Brao to grow up a free man, not as a subject to the atheist Communist Kinh.

Life was more complicated for the village. Besides knowing the Kinh were hunting for them, they also were moving every week or so. It was hard for Chel to be the mother of a newborn and yet constantly move, even with DeAndre doing everything he could to help his wife. But he couldn't breastfeed his child nor could he sleep for her as her nights were broken up into segments so she could feed Brao.

One night, DeAndre risked running out of batteries, but he wanted to know what was going on in the world. By listening to both the Voice of America and the BBC, he got astounding news. Vietnam had invaded Cambodia a few days previous. Though both countries were ostensibly Communist, there were huge cultural and language differences. As the Khmer Rouge continued hunting down anyone suspected of being an intellectual or other enemy, they had begun crossing the border and attacking Vietnamese villages near the border where some Cambodians had taken refuge.

The Hanoi leadership rattled their sabers, but the

Cambodians, possibly feeling the protection of their Chinese allies, kept up the cross-border raids. In February 1979, Chinese troops poured across the border into Vietnam. Just before the invasion, Chinese Vice Premier Deng Xiaoping was overheard telling President Jimmy Carter that "The little child is getting naughty, it's time he got spanked."

But the Chinese hadn't fought a war since 1953, while the Vietnamese had been fighting wars against the French and Americans since 1946. After three weeks and thirty-two thousand casualties, the Chinese declared victory and went home. Because FULRO was a proclaimed enemy of the Vietnamese, the Chinese began supporting the Dega fighters. By the summer of 1979, the leadership at LoMo received both AK-47 rifles and ammunition, then distributed those to the different FULRO villages.

"Now that China is openly in this fight, where are the Americans?" asked Mip one evening as some of the men relaxed while cleaning their weapons. "There were a lot of us who fought with the American Special Forces. I'd say they owe us a return favor."

DeAndre sat silent as the others got themselves wound up about the Americans return. He couldn't be sure, of course, but when he listened to the news from the US, there was always some hint that the country wanted to move on and put Vietnam in the rearview mirror. That would seem like abandonment to the Montagnards. He tried to think of something to say when they inevitably asked what he thought.

"I don't know if they'll come or not. I mean, that's not

something they talk about on the BBC or Voice of America," he said, hoping that would satisfy them.

But the others continued to pepper him for opinions. "Besides, I hate to tell you this, but the mountain area was not all that important to them. It was the South Vietnamese government they were trying to save."

He could tell by their faces that they didn't like to be thought of as "not all that important." He decided to say something else noncommittal. "A lot of Americans know where Saigon is, or Danang, or even Hue, but I seriously doubt if any average American ever heard of Pleiku or Ban Me Thuot."

Still vaguely dissatisfied, the others changed the subject and moved on, but DeAndre felt a slight chill directed towards the only American in the village.

By summer, Chel and Brao were strong and the boy was sleeping through the night. Though he still commanded almost all her time, Chel and DeAndre could now start thinking more about how their new family fit into the village's situation. The village still moved frequently, setting up new camps in the jungle, and the fighters were still attacking the new Kinh villages. They also had to avoid the Vietnamese army, staying away from the roads and disappearing quickly after hitting a village.

The first Chinese arms and ammunition were trickling in. With almost all the ammunition for the M16 expended, it was decided to dispense with those weapons. Life was not dull and it was certainly strenuous, but there was a certain rhythm to it.

DeAndre began working with Ama Tum and Mip

again. He learned that the Kinh had a name for the new villages. They were officially New Economic Zones, made up of both Catholic refugees from the north and southern city dwellers. It was a way of relieving the pressure on Saigon, Danang, Hue, and other cities that had unsustainably large populations due to the refugees from the countryside.

Having realized they needed to protect these settlements, the government both trained the civilians to defend themselves and kept small garrisons of troops in many of them. The Kinh villages had once been easy to attack, but the Dega now had to work together to have enough attackers. Casualties increased. To provide some protection, some other FULRO villages moved across the border into Cambodia. That didn't mean it made them totally safe, but the Kinh were reluctant to pursue them for fear of causing more problems with the Khmer Rouge.

As the 1980s arrived, the village moved again. DeAndre and Chel found themselves in the newest Plei Mrong located in the dense mountain jungles of Mondulkiri Province, Cambodia.

While Chel knew she could not keep her husband from fighting again, she certainly didn't encourage him either. At first, he merely acted as a coordinator, helping put together FULRO attack groups from among the scattered villages in Cambodia and the few still left in Vietnam. He also helped the leadership at Lo Mo plan the attacks. It wasn't particularly dangerous work, but DeAndre did have to travel throughout the mountains which made him more susceptible to attack by enemy troops or aircraft.

By chance, he found there was a side benefit to his work. While on the way home to Chel and Brao, he went through the small Cambodian village of Pu Chrey, which had the usual array of small Mom and Pop stores. Stopping at one to buy some food, he found D Cell batteries—the same type used in his radios. The owner of the shop, though pleasant and polite, wouldn't accept DeAndre's Vietnamese dong as payment, and would only accept either Cambodian riel or US dollars. There was no way DeAndre could get dollars, but the batteries were worth the trek back to the border area to trade out some of his meager supply of dong for riel.

Back with his family, DeAndre turned his radio on and was surprised to learn that a powerful station was broadcasting from Manila, run by the Christian and Missionary Alliance—the same denomination that Smithy and Lee Anne belonged to. Even better, he learned that Smithy would be preaching the following Sunday morning. Almost all the Jarai in Plei Mrong knew Smithy. They all crowded around the radio Sunday morning to listen to their former pastor. They heard him preach in Jarai. Their spirits soared. The prayer room was full that evening as the villagers rejoiced in hearing someone they knew and loved whose preaching filled them with hope.

And DeAndre overheard Mip wonder out loud if Pastor Smithy knew any American leaders well enough to bring the Americans back.

CHAPTER TWENTY

Word came from the FULRO leaders that they were planning a major attack on the Vietnamese and needed Plei Mrong's leadership to attend a meeting. "But you will be going back into Vietnam," pleaded Chel. "Can't someone else go and bring back the information?"

DeAndre understood his wife's concern. At the very least, if the Kinh caught him, he'd be imprisoned and tortured. At the worst, he would be executed after being tortured. It would be like walking into the dragon's mouth. He had hoped the FULRO leadership team would move to Cambodia.

But a few days later, Ama Tum, Mip and DeAndre left their home, walking east towards the border. As usual, they stayed off the roads and moved along jungle paths, arriving safely at LoMo. The day after they arrived, the meeting began.

DeAndre had to brush up on his Rhade, but he soon figured out what the plans were—and they astounded him as well as most of the other FULRO leaders. It was proposed they attack the Dak Mil District Police Station.

His mind began racing as soon as he heard the mission. Were they crazy? The paltry few FULRO fighters against a larger group of well-armed police officers who were ensconced in a solid brick building? Providing they could even survive the attack, how could they escape back to Cambodia? For that matter, why would FULRO even want to attack a large Vietnamese installation?

His last question was answered first. FULRO needed the rest of the world to know they were still fighting. The Dega believed America would help them. Afterall, the Montagnards had fought and bled with the Americans fighting the Communists. The leadership felt that there was a good chance that word on the attack on a police station would put FULRO back in the minds of their erstwhile allies in the United States.

As he listened to the proposed plan of attack, even he began to believe it could work. They would strike during the change of shift around midnight. That assured the smallest number of police on duty as well as the usual minor confusion that takes place anytime there is a shift change. It also meant moving into position under cover of darkness. As to escape, the plan was to go north rather than west back towards Cambodia. Once the escapees were deep in the jungle, they would hunker down for a few days so as not to expose themselves by moving. DeAndre began to feel better.

Then the suggestions for improving the plan began. Knowing they would need more ammunition than the normal amount used attacking a newly built Kinh village, it was decided to send in a few people beforehand and leave munitions in a cache for the fighters to pick up just prior to the attack. The talk became more exciting as the

FULRO leaders began to believe the attack could succeed.

The confab continued for two more days. DeAndre began to realize he was not the only wise combat veteran—these fighters knew what they were doing. They even set up a makeshift sand map, plotting out movements and times on it.

The three men from Plei Mrong returned with lighter hearts than they'd had on the way to LoMo. Each village had provided an approximation of the number of fighters it could produce (being sure to leave some behind to protect the village). DeAndre was astounded to hear that FULRO's attacking force would number almost a thousand well-armed, well led, and confident warriors.

The attack was set for a Saturday night when it was guessed that some police would be drunk. A new moon would help the Dega as they crept into position, prompting Ama Tum to instruct everyone to wear dark clothing covering all their bodies, meaning no loincloths. Slings were removed from their Ak-47s to lessen the chances of a tell-tale rattle. Extra ammunition could only be draped across the chest if it was covered with dark cloth to reduce reflections.

For the first time, DeAndre would participate in the attack rather than be hidden. Chel didn't like the idea, but she supported her husband. The leadership also thought that it wouldn't hurt if the Vietnamese knew there was an American attacking the station. The Vietnamese authorities were more apt to point that out, increasing the chances of it becoming a story worthy of western news media.

Well before the attack, the first ammo bearers left the various FULRO villages to create hidden stockpiles of ammunition and grenades near the police station. The leaders decided to attack from the south side of the small town to avoid having to cross or go around a small lake.

The fighters from Plei Mrong left four days before the scheduled assault, stopping by their cache before getting into position. They waited quietly through the daylight hours, hidden in a small grove of trees. At nightfall, they quietly began their preparation by applying ashes to their faces and hands. They could hear raucous laughter coming from parts of Dak Mil—Saturday night was in full force, and the beer was flowing freely. The shift change would happen at 11pm, meaning they had to be in place by 10:30.

By 10:45, members of the new shift began to report to their duty stations and hold brief conversations with the outgoing people. The FULRO fighters had done their homework and knew where each duty station was located, and appropriately sized teams were assigned to each. At 11pm sharp, Ama Tum pulled the trigger, signaling the rest of the force to open fire. Ama Tum's target was the shift commander, who was dead before hitting the floor.

Mip was assigned the booking desk for the jail. The first member of his team burst through the outside door, felling the first uniform he saw. A split second afterwards, his other team members took down the remaining jailers. In other parts of the station, a hand grenade thrown into a crowd of Kinh preceded the shooting.

Some of the teams met return gunfire when the new

shift personnel showed up a little late, heard the shooting, and had time to assess the situation before engaging the Degas. Not all police are armed while on duty, and those fled. The few that either were armed or could get weapons quickly, put up a pretty good fight against the numerically superior Degas. Unfortunately, some were wounded. Fewer killed.

DeAndre was on a team assigned to an office on the second floor. Of course, they could not get there unseen, so they had to wait until the shooting started, then bolt up the stairs. But that few seconds gave the outgoing shift time to open the gun cabinet and distribute some of the rifles. He heard the bolts on the AK-47s slam shut just as he was about to round a corner. He stooped as low as he could get, then poked his head and weapon around the corner. As enemy bullets whizzed by over his head, DeAndre kept his cool and dropped one policeman who had not spotted him yet.

Ducking back behind the corner, he waited a few seconds until he heard shots he knew were from his comrades. Knowing the enemy would now be distracted, he came out in the open and shot two more policemen. By this time, the police ranks had been decimated and return fire all but disappeared.

The planning had even included ways to evacuate their wounded. As DeAndre and others covered them, three teams tended to the two wounded and one KIA, quickly moving them out of the office and down the steps. DeAndre fired on some police who saw the evacuation teams, but didn't know how effective he was.

The evacuation teams headed out into the darkness while DeAndre and the remainder of the team headed for a rendezvous point. There was still sporadic fire coming from the police, so DeAndre kept as low as possible while allowing him to run.

He didn't run quite low enough. He felt a sharp stinging sensation in his upper left arm. He knew he was hit, but a quick look told him it wasn't bad. There was a brief wait at the rendezvous point before they all headed north to the escape path. One of the other fighters had enough time to dab the wound with disinfectant, then apply a bandage—one that was left over from the old Special Forces days.

Moving as fast as possible at night, the force moved well into the jungle north of Dak Mil, and at dawn hunkered down in the thickest jungle they could find. Again, good planning had made a difference. They saw numerous helicopters and airplanes flying over the area to the west of Dak Mil, on the most direct route to Cambodia. Their joy in the mission's success was tempered by the knowledge that some of their number had died. They wouldn't know how many and the names of the fallen until they got back home.

As planned, the FULRO fighters did not return to LoMo, but rather to their own camps. It took a few days, but when DeAndre and the others entered Plei Mrong, they noticed that some of their own had not returned. Nobody knew if they were dead or wounded. Two agonizing days later, a runner from another camp came with the casualty figures. Three of their own were now dead, their bodies in LoMo. Another six wounded, one of them badly.

There was much sorrow, crying and wailing in the village as a team set out for LoMo to bring back the bodies as well as any wounded capable of making the trek. "Yes, we won the fight," DeAndre told his wife. "But there are many more Kinh than there are Dega."

Chel wasn't quite sure she understood and asked her husband to explain.

"There are millions of Kinh in Vietnam, but only a few thousand of us tribal people. At Dak Mil, we had twelve men killed. I don't know how many Kinh police were killed, but let's say fifty. There are more than enough Vietnamese to replace their dead but replacing twelve of us will be a lot harder."

He didn't know if she understood the math, but he went further saying, "Our dead were a much larger percentage of the population for us than the Kinh." He didn't mention the three recent widows in Plei Mrong, but Chel knew.

There was no mention of the attack on the BBC news or Radio Free Asia. Even the Voice of America barely mentioned it at the end of the news. "Yesterday, a group of assailants attacked a rural police station in western Vietnam. It is unknown how many members of the assaulting group were killed, but twenty-two police officers died. Authorities say they are pursuing the terrorists."

All the people of Plei Mrong were deeply disappointed. They had pinned their hopes on having some serious media attention paid to the attack, but only one news source had mentioned it. The Sunday church broadcast from Manila was equally silent.

A few weeks later, the LoMo leadership decided to move to Cambodia as well. That left no organized FULRO fighters inside Vietnam.

Like the people of Plei Mrong, those of LoMo decided to keep the name of their village too. As it happened, it was the closest to DeAndre's village, making it much easier for the leadership teams to talk to each other.

FULRO had another problem. The Khmer Rouge no longer brought weapons and ammunition to them but let it be known they were still supportive. The Dega would have to send people to China to pick up weapons and ammunition. That meant still another decrease in the number of fighters.

To compound all this dark news, the Khmer Rouge were rapidly turning from friends to enemies. The Vietnamese invasion of Cambodia in 1979 was the beginning of the end with the Vietnamese army overpowering the smaller and poorly equipped Khmer Rouge. Worse, they began to fight among themselves and with the Chinese.

Plei Mrong wasn't the only FULRO village that had the problem of Vietnamese strength. They all did. Now that there were no FULRO fights inside Vietnam, there were fewer and smaller raids, mostly road ambushes of civilian trucks and pot shots at Kinh farmers working on the coffee plantations.

It took a while for the village to mourn the death of their warriors who had died in the attack on Dak Mil. Besides the three who had died immediately, one of the wounded had infections that the Dega had no medicine for. The lack of ammunition reduced the number of raids they could participate in. The number of widows and fatherless children meant other men had to step up to do some of the farming and hunting the dead men would have done.

But there was joy in DeAndre's life. He and Chel were happy watching their son grow. A rambunctious toddler, he got into everything and became a beloved annoyance throughout the village. He was too young to know it, but he was a child in a perfect home: his parents loved each other, giving him all the love and security he needed.

CHAPTER TWENTY-ONE

Not all Americans ignored the Dega. First Sergeant Eric Wozniak had retired from the 82nd Airborne Division at Fort Bragg, North Carolina. He and his wife Heather decided to settle down in the beautiful Piedmont area surrounding Asheboro, North Carolina. Their kids were grown and gone, and they decided on a life near the Blue Ridge range. Eric remembered the mountains near Kon-Tum and Pleiku with nostalgia.

Even though she couldn't identify with her husband's memories of the Vietnamese peaks, she did like the green splendor of the area. The hour and a half drive to Fort Bragg made using the medical facilities, post exchange, and other services fairly easy, yet was far enough away to not be an "Army town" like Fayetteville.

Once in a while, Wozniak wondered what had happened to his former boss in Kon Tum. The two had clicked very well. The Army no longer had a record of him beyond saying he was, "Missing in Action." After a few years, he did a bit of searching and found Mrs. Washington's address, but there was no phone number. A bit more sleuthing revealed the name of a little Baptist church that was fairly near the address. A phone call there put him in touch with Pastor Willis, who quickly affirmed that he knew DeAndre and Mrs. Washington.

"From what I understand, the Army isn't totally sure what happened to Captain Washington," said the pastor. "The Survivor's Assistance Officer that I accompanied to see Mrs. Washington seemed to think he was still alive, and that most likely, he was captured. You might ask some other folks in the Army if anything else is known."

That piqued Eric's curiosity. How could Captain Washington get captured if he was flying a desk in Pleiku? Remembering that his former boss had spent a fair amount of time in Kon Tum working with an American missionary named Smith and knowing that most missionaries to the Montagnards were part of the Christian and Missionary Alliance denomination, some more research revealed that the Reverend Taylor Smith was living near Raleigh.

After the pleasant phone call, the Wozniaks drove over to visit the Smiths. Lee Anne cooked a fabulous dinner, and the conversation was about Vietnam in general and DeAndre in particular.

And there was much to say about their mutual friend. First, the topic of Chel. The retired NCO soon found that not only had Captain Washington stayed in Vietnam with the Jarai, but he had married Chel. Next, Smithy also talked about the ruse that made it look like DeAndre had been captured. DeAndre's driver was convinced that the "capture" was real. But Smithy didn't know about anything that happened after he and Lee Anne had left Pleiku for Saigon.

After making sergeant Wozniak swear to secrecy about DeAndre, they left, the two couples vowing to keep each other up to date.

CHAPTER TWENTY-TWO

In 1980, life in Plei Mrong had only gotten more difficult, even after the move to Mondulkiri Province in Cambodia. Combat with the Kinh since the fall of Saigon in 1975 had decimated the ranks of the fighters. Of course, that also meant there were fewer men to farm. Once their friends, the Khmer Rouge were now a threat.

It was decided to merge with another FULRO village for their mutual benefit. The leaders of Plei Mrong and Plei Doch met and decided to occupy the area of Plei Mrong, but they would also adopt a new village name: Plei Mook Trê. They were about seventy miles from the border with Vietnam, which was good in that it would be very hard for the Vietnamese army to attack them, but it also made it harder for them to attack the Kinh villages.

FULRO once boasted an army of over 10,000 soldiers, but by 1980, they were down to a mere 2,000 fighters. Nor did they have enough weapons and ammunition. Their sources had dried up. By 1982, they had a new problem. The Khmer Rouge had all but been eliminated and Cambodia was governed by the United Nations Transitional Authority in Cambodia (UNTAC.) Worried about their own Montagnard minority, the new regime in the capital of Phnom Penh told UNTAC that FUL-

RO would either have to be disarmed or the army would attack it. UNTAC was charged with the responsibility of ridding the country of all foreign military groups—and FULRO was a foreign force. UNTAC also knew that if the Degas were sent back to Vietnam, they would be severely persecuted. The United Nations High Commission on Refugees (UNHCR) would love to have granted them refugee status, but that would open the floodgates to millions of Vietnamese people in Cambodia.

It was now seven years since Vietnam had been united under Communist rule, and FULRO was too weak to execute major attacks. It was under these depressing conditions that the village of Plei Mook Trê sent out a small patrol, under DeAndre's leadership, to find targets of opportunity inside Vietnam.

DeAndre steered his ten men away from Route 14 as it was too close to the border where the Vietnamese expected attacks and patrolled aggressively. Instead, DeAndre found a nice curve in the road leading from Dak Mil into Ban Me Thuot. It offered deep jungle for cover and a long curve that allowed the Dega to rake the flanks of any vehicles that passed. Escape to the north would not be what the Kinh would expect.

The men had brought a lot of food with them, mostly in the form of cloth tubes filled with rice that were slung around their necks. They had the village's only automatic weapon, an old Chinese made RPD. There were only two ammunition drums which gave them the ability to put 200 bullets on target quickly. The remainder of the patrol members carried AK-47s. There were no hand grenades.

Well hidden, the patrol waited for either a single truck or small convoy. They were a little wary hitting just one vehicle as it might be bait and had a larger force behind it to take out any ambushers. A day went by before the lookout spotted three Ural trucks with a UAZ jeep-like vehicle leading the way. The soft sides of the truck were rolled up and all that could be seen were some wooden crates. Each vehicle had one soldier riding "shotgun" next to the driver, with one truck sporting a mounted Type 667 machine gun.

With limited firepower, DeAndre didn't want to start shooting too soon. The truck with the machinegun needed to be in close range so that three of his fighters could take out the gunner before he had a chance to open fire on them. In addition to the driver and guard, the UAZ also had a large radio in it, along with an operator. DeAndre designated two of his men to take out the radio operator. The remainder of the fighters were assigned targets and DeAndre would fire the opening round.

The driver and guard on the radio vehicle were enjoying a lively conversation and laughing out loud when DeAndre's bullet hit the operator's head. Instantly, the machine gunner slumped over his gun before he had a chance to pivot it around. Neither did the driver of the UAZ have a chance to pick up the handset. The only rounds fired by the Vietnamese soldiers came from the guard of the first truck and those were fired wildly in the wrong direction. The ambush was executed in textbook fashion: quickly with great violence.

It was not safe to administer a coup de gras because

it would consume too much time, but it was also obvious that all the enemy were dead or seriously wounded. Likewise, it wasn't safe to take ammunition or weapons.

DeAndre's cease fire command was quickly followed by all of them rising from their positions and heading out to the north through the jungle. Traveling as only Montagnards can move through the dense trees and vines, the attackers were long gone before Vietnamese forces arrived on the scene. Nonetheless, DeAndre kept them moving until deep dusk, giving themselves just enough light to find and settle into a defensible position.

Even with the dawn, the patrol kept moving north until noontime when DeAndre changed the direction towards the west and Cambodia. They were super cautious as they crossed Route 14 as that would be a great place for the ambushers to be ambushed.

Being only a short distance from the road to the border, they moved quickly and tried to get back into thick jungle. The Vietnamese had cleared much of the area to make it more difficult for FULRO to cross into Vietnam, but that's all it did. They made it harder but certainly didn't prevent any FULRO fighters from returning to Cambodia.

On the third day after the ambush in Mondulkiri Province, the lead man paused, looking down the trail. A figure stood motionless. DeAndre's combat instincts alerted. Scanning the surrounding jungle, it took time before he allowed himself to breathe easier. Alone? The man was alone? DeAndre chewed on that, asking himself why the man would be standing out in the open as if wanting to be found. He finally decided the man was safe.

CHAPTER TWENTY-THREE

Accompanied only by the lead man while the rest of the men stayed crouched and vigilant, DeAndre cautiously approached the man. He said his name was Rahma Dock, a Bahnar from the Pleiku area who also spoke Rhade and English. He was neither lost nor hungry. In fact, he was hoping to run into the patrol.

He'd found Plei Mook Trê a few days earlier when he heard people talking as he walked on a nearby trail. While in the village, the people gathered around him, curious about the picture he painted of the possibilities beyond the Cambodian jungle. "I made a deal to leave your village, but had them promise I'd be allowed back."

Looking at DeAndre, he wasn't sure he was believed. He went on, "You can find safety in Thailand," he urged the American, "and there's a chance the Montagnards can start anew in America."

When the villagers told him that a patrol from the village had crossed into Vietnam, led by an American, "I just had to come find you, sir. I couldn't sit idly by in the village, waiting for your return." This convinced DeAndre that Dock was legitimate, and the patrol continued on towards home, with Dock in tow.

Once back in Plei Mook Trê, Dock told his tale to everyone. He said, "I was an administrator for the southern government in Pleiku, and I worked with the Americans. When the Communists took over, I was arrested. They said I was a traitor and I was sentenced to three years in prison. They made me do hard work in the rice fields."

As chance would have it, one day he and some others saw an opportunity and escaped into the nearby jungle. He soon joined a group of rebels, attacking Kinh villages, ambushing vehicles and raiding small military outposts. But, as the Vietnamese wore them down, Dock and the others retreated to Cambodia where they ran into a large nasty group of Khmer Rouge who imprisoned him along with 250 other FULRO.

Then came another break. The Vietnamese heavily bombed the Khmer Rouge encampment and the Dega were able to flee into the jungle. Without maps or weapons, they finally made it to the border with Thailand. There they talked to Thai Lieutenant General Chavalit Yongchaiyudh. He advised the Montagnards that the Americans were no longer interested in fighting the Vietnamese. He also recommended they apply for asylum to the United Nations High Commission for Refugees. When it was granted, they were placed in a refugee camp in Thailand.

Someone in the Highlander group in the camp knew Jim Lienghot, an acculturated Montagnard living in the Los Angeles area. Lienghot knew Donald Scott, a former American medical administrator in the Central Highlands. Scott went to work spreading the word to Vietnam veterans' organizations and groups of former Green Berets that the highlander former fighters were languishing in the Thai refugee camp.

Dock urged the people of Plei Mook Trê to go with him to Thailand, but that it would be a long and hard journey, especially for families with babies and small children. Most of the fighters still believed the US would help them.

Nay Moul had been with a Mike Force team, which was trained and led by American Special Forces. He spoke for the other men who had served with the Americans. "They are my brothers," he said. "They will always be there to help and fight with us. We fought and bled together."

Nobody disputed that idea. Then Ama Tum offered an idea. He would lead a small delegation to Thailand, get more information, then return to the village for a final decision. Only a few fighters would accompany Ama Tum while the rest would stay put. It was decided DeAndre would not go. Perhaps having an American in FULRO would pose a problem in seeking asylum.

Ama Tum's return to Plei Mook Trê was out of character for a Jarai man. He was over the top with exuberance. After jumping up and down while hugging his wife, the two went to see Chel and her husband. "DeAndre! This is it!" His excitement flowed over his face. "We talked to some Americans who say they will help us get something that we need before we go to America."

"What thing?" asked DeAndre. "I dunno—sound something like sigh-lum."

It took a few minutes to explain asylum to Ama Tum.

The Montagnards had no such thing simply because they were not an actual country.

DeAndre also asked who these Americans were, only to find out they were from the American embassy in Bangkok. They even showed Ama Tum some small papers with a picture of the person. DeAndre figured out that was the Americans' credentials. DeAndre asked some more pointed questions to be sure Ama Tum had really met Americans.

Ama Tum directed that a village meeting take place the next morning immediately after morning prayers. "Father God, we need your guidance. Put it in our hearts if we should go to America" intoned the worship leader. The congregation all began praying at once to find what God wanted them to do. Rahma Dock had never seen such a fevered group of Christians, and he was impressed by their sincerity. He had never become a Christian and still believed in the spirits.

The congregation got their answer quickly. After the final "Amen" was said, the villagers grabbed something to eat. They were all eager for the village meeting to start.

Though village meetings had few formalities, the meeting reached a decision within fifteen minutes. They would go to Thailand. Each knew a lot of planning was required—the women how to move the children and household stuff, how to prepare meals on the way, and how to keep the little ones quiet. The men had to make tactical plans. How would they scout ahead to avoid the Khmer Rouge? How to provide flank security? How would they provide security at night?

When the meeting ended, DeAndre and Chel returned to their sleeping area. "I have one more thing to prepare for trip," said Chel. "I am going to have another baby." DeAndre's head spun. He hugged his wife, saying, "I love you" over and over, while his mind raced on saying to himself that if he had known this before the meeting, he would not have been so eager to go. Brao was old enough to make the trip, but Chel would tire easily. He wanted to ask why she hadn't told him earlier, but he'd been a husband long enough to know better.

Though people didn't have the privacy they once enjoyed, couples still felt comfortable enough to make love, as long as they were quiet. Chel took the opportunity to signal she wanted her husband. "Are you sure? " he demurred. "You're pregnant."

"But not so much that it hurt baby" replied Chel. She talked no more, but her hands let him know what she wanted.

It would take a couple of days to prepare for the long walk. Backpacks were loaded with food, cooked rice stuffed into socks that doubled as food sacks, and the women talked about food that would not need cooking. The smell of the fire and the cooking food might give away their location.

The men developed an "order of march" that depicted where each family would walk, including which man would precede the main body, acting as the point man. A schedule was developed that rotated which men would move along the flanks on a rotating schedule so that everyone was with their family at least once a day. A system of listening posts was arranged for the nights.

Dawn broke, and the people of Plei Mook Trê met for prayer, a quick breakfast, then they began the long trek to the Thai border.

Far from being handicapped by her pregnancy, Chel rose to being the motivator and encourager for the women. She had spoken long and hard with her son, telling him he would have adult responsibilities. She expected Brao to be a leader for the younger boys, and he rose to the challenge. Chel carried as much as any woman, which was a lot. The men tried to take some of the load off their wives, and the few single men helped the families, but carrying the loads for days on end was exhausting.

Ten days passed. Doch told Ama-Tum that they were about halfway to the border. As they trudged along, he watched the women and the children, some of them struggling. Finding DeAndre, the two pondered stopping for a day to let the people rest. The two knew many of the women were widows, and while some of them had older teenagers assigned to help them, some folks seemed to be getting overwhelmed.

That night, after evening prayer, Ama Tum pronounced the decision—the villagers would take the next day off. Many of the widows protested—they didn't want to be the reason for delaying the villagers getting to Thailand and safety.

Doch stepped in and said they would soon be in the area of Tonle Sap, the big lake fed by the Mekong River. That area was more inhabited than the areas they had traversed so far. Everyone would need to be well rested so as to be able to flee if discovered. That settled it, though some of the wives were still a bit disgruntled.

Sure enough, three days later, they crossed more and more roads, saw tilled fields and other indications of settlements. Sometimes, the leadership asked everyone to crouch while moving to avoid being seen by farmers. Being doubled over made walking a lot more difficult.

After two days, they were back in thick jungle again, and all felt relieved. They all knew without being told that they were getting closer and a certain fresh energy became palpable.

CHAPTER TWENTY-FOUR

DeAndre's western ways still clung to him in some ways. The Montagnards had no calendars. They reckoned time by seasons. But DeAndre knew they were now in 1986. That was confirmed when Thai border guards appeared, and the group was led to a camp. Safe at last. The weariness overcame them all, and they collapsed into heaps. The camp, run by a volunteer group made up of both Thais and Americans operating out of the American embassy, provided warm food, good, sheltered sleeping arrangements in tents, a place to wash clothes, and safety. The men were asked to turn in their weapons. They knew this was coming and understood the reasons. There were no complaints as Thai soldiers took them.

The villagers settled in for a couple of hours, then gathered for a worship service. Obviously, they were not the first Christian Dega to come through the camp as the volunteer group even provided a small gathering place marked by a cross.

Most of the men had taken showers before when working with American soldiers and knew what to expect, but their wives did not. The volunteers smilingly looked away as husbands and wives shared a shower, producing girlish giggles.

Morning came and they were all refreshed. After morning prayers and breakfast, a group appeared from the American embassy. They were surprised to see a tall black man in the group. DeAndre identified himself as American, and while they believed him, he still would need to go to the embassy in Bangkok. The rest were told all about asylum, and the process of going to the United States. Chel was to stay with her husband as she would be allowed to immigrate by a different means than asylum since she was DeAndre's legal wife.

But they had a problem. One does not walk through the streets of Bangkok dressed in mountain togs. The volunteers had no problem finding clothes for Chel, but big, tall DeAndre was more of a problem. It was decided to go to Bangkok where embassy personnel could help them.

Two days later, two men from the American embassy came to take DeAndre and Chel to the American embassy in Bangkok while Brao would be watched by one of the Jarai women. They were a bit startled to see DeAndre in Jarai clothes but understood after one of the volunteers explained the situation. Money had been donated by embassy staff when word came down that an American was dressed only in native garb from the mountains.

The trip to Bangkok was in the back seat of a 1979 Chevrolet sedan. Chel had never been in an automobile before and snuggled up against her husband in the back seat. DeAndre smiled as he put his arm around her. "I am a little scared," she said. "Go very fast." DeAndre understood then. He felt her tense up as they went over bumps.

"No doubt about you, Captain. You were listed as

Missing in Action, and your personal information was made available by the Department of Defense" intoned the pleasant man at the embassy. Mr. Jacobsen went on to say, "Obviously, you are a U. S. citizen. Next, we'll need to gather information about Ms. Yah here." DeAndre was mildly startled to hear Chel's family name. "And we'll need to find the minister who married you and see if he has the proper credentials. You told us that Rev. Taylor Smith is part of the Christian and Missionary Alliance denomination. Should be easy."

There was a knock on the office door, and a clerk came in to give the man a message. "Excuse me, but I'll be right back." DeAndre looked around and saw the room was the usual boring off-white that most government buildings are painted with, but there was also a nice family photo on the man's desk, showing his wife and three kids.

"Sorry about that, Captain." He went on to explain that Chel would not become a US citizen immediately but rather be given a visa which would qualify her for a so-called "green card" that signified a permanent resident. It would take a while before she became a citizen. Jacobsen went on to say that he and Chel would stay at the embassy until their departure for the U.S. "I'll be going with the Degas, I hope."

"Oh yes. You're their titular leader," was the smiling reply.

That's when DeAndre noticed his wife seemed to be in pain. Mr. Jacobsen noticed too. "Chel, you okay?"

"I think so," but just as she said that she winced again. "Mr. Jacobsen—as you can see, my wife is well along in her pregnancy." The administrator called in another staff member, and the two huddled together for a moment before she left the room, only to return very quickly. "It's all set. Bumrungrad expects her within the hour, and the OB-GYN on call will handle it."

"What's Bumrungrad?" asked DeAndre. "It's the big international hospital here. We have a contract with them to provide medical services to staff and in-country Americans. That's you two. Their ambulance will be here shortly to pick up both of you. I think it's better if they come up here for her rather than risk her walking downstairs."

Chel needed all the assurance her husband had given her. She'd never seen a doctor before, much less been in a hospital, but everyone she met was kind and courteous. As she was carried inside, a lady in a long white coat met them, introducing herself as Dr. Zolinski. Her smile was warm and reassuring. DeAndre said he would translate.

"Are you her husband?" "Yes. And I know we don't look alike," he said as he smiled back. "By the way, this is the first time she's been in a hospital and the first time she's ever seen a doctor."

After thanking DeAndre for that information, she asked if there was anything else she needed to know. "Yes. We just finished a long walk from the middle of Cambodia. My wife carried just as much as the other women and still took care of our little boy."

Dr. Zolinski thanked him for that information as well as finding out that Chel had been through a delivery once before—and probably without pain killers. "But we still need you to translate. Will she mind if you stay with her to translate?" DeAndre posed that question to his wife. "She's fine with that, doctor."

The examination didn't take long. "She had carried a lot of weight while walking for a couple of weeks. She just bruised a muscle in the area near her uterus," explained Dr. Zolinski. "There's absolutely nothing for her to worry about. I've given her a muscle relaxer. Here's more," she said as she handed him the prescription bottle.

An hour later, Chel felt fine and another American car took them back to the embassy. Mr. Jacobsen and his team had prepared a small two room suite with an adjacent bath for the three. Chel was excited seeing this would be her "home" for a few months while all the paperwork was done to go to America. Brao was downright over the top with enthusiasm when he learned he had his own room and bed.

The details approached being overwhelming. The American asylum system is based on being able to properly vet applicants. But, since the Dega were not a nation and never had been and had no actual government (except the inimical Vietnamese), there was no way to determine if someone was a criminal, wished harm on the United States, or had a violent nature. Since few had ever seen doctors, there was no way to know if they had long term infectious diseases. Most of the screening was done by talking to former Special Forces soldiers who had trained and fought with them during the war.

But some positive things happened while they were waiting. Old maladies had been cured or healed with the help of modern medicine. Chel received her green card within a month. More mountain people began wearing western clothing and some learned at least a smattering of English.

As days grew to weeks that grew to months, boredom set in. Having seen this before with previous asylum seekers, the American staff began showing them more parts of American culture. Baseball was popular, especially since they had at least one Jarai who knew all about the game. "It's two and one, not one and two. Balls always come first, then strikes." As he taught the men how to bat, DeAndre looked forward to doing that with his own son. He and Chel spent three days back at the embassy during the middle of the week so they could help with the asylum process.

As he was exhorting one of the men to touch the base rather than just running by it, he looked up into the stands. One of the village women was bending over Chel. Soon, a couple of the men were helping carry her to a nice grassy area. The embassy staff called for an ambulance to take her and DeAndre to Bumrungrad in Bangkok.

Two hours later, DeAndre became a father for the second time. The new arrival was a healthy girl. Mom and Dad had already picked out a name. She would be known as Mary. After the nurse had cleaned her up and she was nestled in her mother's arms, DeAndre came into the delivery room. He would wait until Chel was in her regular room before he held his new child.

And now they were four.

CHAPTER TWENTY-FIVE

Six months later, an American Airlines plane carrying the usual load of business people, tourists, and leisure travelers, landed at Piedmont Triad International Airport near Greensboro, North Carolina. Most of their conversation had been about the oddly dressed darker skinned foreigners onboard. DeAndre told his people to stay in their seats until everyone else had left, then they calmly departed through the jetway, entering another strange airport.

Many of the men wore jeans while others had on shorts. T-shirts were the norm, and sandals—the shower shoe variety—on all of them. There were many instances of the clothes fitting poorly. The women also wore jeans, t-shirts, and sandals but many sported colorful beaded necklaces.

Both genders carried small children, while older kids walked on their own. All were bone tired. They'd never been on airplanes before, and most were fearful, but had boarded the airplane at Bangkok's airport, bound for Tokyo. DeAndre had talked to them at length about flying, telling them it would be noisy when taking off, how to swallow when feeling pressure in the ears, how to use the bathrooms, and why they needed to keep your seatbelt fastened.

After flying six long hours to Tokyo's Narita airport, they walked to the departure lounge for their American Airlines flight to Dallas, where they waited for five hours. DeAndre used some of the money the embassy staff had given him to buy food for the travelers. He also warned them that the flight time would be much longer—about 14 hours.

The newness had worn off the children and they were getting restless, but the women managed to keep them from being too boisterous. DeAndre encouraged them to walk around the huge airport, but to remember how to get back. He had them memorize the sign that said Gate 26 and say, "Where Gate 26?"

The Dallas-Ft. Worth airport was equally intimidating to the Highlanders, but they were so tired, they didn't have the energy to fuss. This time the wait was much shorter and 90 minutes after landing, they were on their flight to Greensboro.

They walked as a group down the hallway towards the airport lobby. As they went through the doors, they were greeted with applause and cries of "Welcome" in their native language. Some of the Montagnard men began to recognize American soldiers they had once worked with. The former Special Forces men remembered some of the Jarai or Rhade languages they had once learned, doing their best to make the newcomers feel welcome.

After a noisy fifteen minutes, DeAndre was introduced to Congressman Frank Wolf of Virginia who had sponsored the Montagnard Refugee Assistance Act that brought the Montagnards to the U.S. DeAndre wanted the congressman to say a few words, but Wolf declined, saying the day belonged to the asylum seekers.

But they couldn't continue clogging up the airport lobby, so the group filed outside to board buses that had been hired for the occasion. Nobody had any checked luggage. DeAndre got an aisle seat, and much to Chel's surprise, the opposite aisle seat was taken almost immediately by a man who introduced himself as Ben Cooper. "I knew some of these men. Back in the day," he smiled, "I was Special Forces assigned to a team in the Pleiku area. Like all the other Vietnam vets here, we had a lot of respect for our Yard friends. We've gotten together to form a nonprofit called 'Save the Montagnard People.'"

"Pleasure to meet you, Ben. I'm DeAndre Washington. If you were around Pleiku, do you speak any Jarai?" With that, Ben answered in halting Jarai, saying it had been such a long time since he'd used the language that he was pretty rusty.

"We don't use their native languages on purpose. They need to learn English, and they will learn better if they are immersed. And, like everyone else here, I know you. You're the POW who somehow managed to escape. You know you are still considered on active duty since you were a POW."

This bit of information made DeAndre's head spin. All those years in the jungle and he'd almost forgotten about his status. America was about to treat him like a hero—as a POW—but in reality, he was a deserter. He would have to tell Chel to be super quiet about that, even if someone asked.

The chit-chat continued until the caravan of buses arrived at the destination and turned up a small private road. Looking out their windows, the people of the Highlands saw a familiar sight—a small village of longhouses. Some were Jarai, others Rhade, Katu, and other tribes, but they were all built like the longhouses they had left in the mountains of Vietnam. The buses exploded in chatter, fingers pointing at the familiar buildings.

Tired as they were, their excitement overcame their weariness. Home—this was their new home, but in a place where they didn't have to worry about being attacked.

As the buses came to a stop, the new arrivals erupted in new cheers as they saw Montagnards coming out of the long houses to greet them. Jarai sought out Jarai and Rhade found Rhade, so that everyone could speak their tribal language for a little while until everyone went back to English. Each incoming family would have a mentoring family, even for the children. This was all an effort to make the new folks feel at home and welcomed.

The guides then took their charges to longhouses, where they were assigned rooms. The two families would be close to one another so as to make it easier for moms, dads, and kids alike to get some help.

When DeAndre was asked if he wanted to be in an American style home or a longhouse, he quickly answered that he wanted his wife to be comfortable so they opted for the longhouse. Their hosts had provided all the essentials of home, including two extra sets of clothing for husband, wife, and any kids.

But the longhouses weren't quite like the longhouses of Vietnam. For one thing, they had electricity. DeAndre was surprised and delighted to see a bare bulb hanging in their room. "There are outlets along the outside wall. You will be able to plug in whatever you wish," said their guide.

As DeAndre, Chel, and the kids checked out the room, they engaged in hushed but excited conversation as they found things that were traditional and others that were new to Chel. DeAndre took time to carefully explain the room's small refrigerator and what it could be used for. Another woman who lived in the longhouse came in with a beautiful new blanket, woven with familiar Jarai patterns.

Reverting to English she said, "Later in year, it cold here. You need this." As the "Thank yous" followed her out of the room, another lady came in carrying more clothes for Chel, Brao, and Mary. She asked DeAndre if his wife wanted Dega clothes or western ones. He didn't know. He vowed to ask one of the men in Save the Montagnard People what she should wear.

There was a fire pit outside that Chel could use for cooking, but she'd also found a two-burner countertop stove in their room. As the two talked about it, they heard a timid knock on the frame of the room door. Turning, DeAndre saw his mother.

DeAndre had pondered this moment for years. He'd dwelled on it while flying across the Pacific. In quiet moments with his wife sleeping next to him, he thought of how he would react to seeing his mother again.

All such planning deserted his head the moment he saw her. Without any hesitation, he took only two steps to be at her side and picked her up as he hugged her, all the while whispering, "Mama, I love you!"

But Jada Washington outdid her son in her emotion. Returning his hug, she shouted, "He's home! He's here! My DeAndre is back home!" The two sobbed at once, all the while hugging and stepping back to look at the other. Of course, Chel didn't need to be told who this chocolate brown woman was—she knew it was his mother. She smiled as she saw part of her husband's family for the first time.

When, after more hugs, they stepped back to look at each other again, Mrs. Washington looked over his shoulder and smiled, "Come. Introduce me to my new daughter-in-law."

Chel saw the two of them turn towards her. She smiled at them, but she also smiled for herself. She was proud of her husband, yet she was proud to be wearing a beautiful Jarai dress. Stepping from his mother's side to his wife's, he said simply, "Mother, this is Chel, my wife." Surprisingly to Chel, Jada stepped forward and hugged her. Chel may not have been used to the public display of affection used by Americans, but she totally understood she was accepted, totally and completely, without reservation.

Without planning it, both women moved to DeAndre's side, both putting their arms around him for a hug. Pastor Willis beamed from the doorway.

Brao came into the room, leading his toddler sister by the hand. "Oh, how precious! I am a grandma! What are their names?" Chel walked over to her now-shy son and daughter and said "This Brao. This Mary. They our children." The heavily accented English startled Jada, but she quickly recovered and gave DeAndre's wife another hug. "Good English," she smiled.

DeAndre stepped in, and speaking to his mother, told her that Chel is very bright and he had taught her some English, but not much because there was no need for it in the jungle.

"I want to help her learn, DeAndre."

"You most certainly can. All these people will be taught English, including Chel. You can help her practice."

"Oh yes, and I will teach her the words she needs in my kitchen."

The reunion continued for an hour, Jada being sure she included Chel in the conversation and worked with DeAndre to use as much English as possible. Of course, she could not understand the language they were using, but DeAndre was good about translating it for her.

Ben Cooper came in and quietly told DeAndre that all the newcomers needed to meet outside the longhouses to start their orientation in America. He also told DeAndre that he and Pastor Willis had done some planning. Since DeAndre and Chel were the only people with American families, the two had arranged for a car to come pick up the couple the next Saturday and take them to Mrs. Washington's house to spend the weekend there.

"Is that okay, DeAndre?" asked Pastor Willis. "We don't want to go over your head." "Are you kidding me? Chel spending the night at her mother-in-law's house and attending worship service the next morning? Perfect!" beamed DeAndre.

The next week was exhausting. The newcomers were assigned an American to intervene on their behalf for such things as where they would worship and if there were any Jarai bibles. They wanted to know if they would learn to drive. One mother had a child with a bad and worsening cough and wanted medical care. They had two different English classes: one to learn the language and the other to learn the culture. Since the children had never been to school before, the parents had to set that up—and school explained to the children.

The busy week over, Pastor Willis himself came on Saturday to pick them up and take them to Jada Washington's home. During the drive, DeAndre told Chel that this was the place he'd grown up, how much he had loved his father, and what a happy family it had been. After a Southern style welcome of hugs and fussing, Jada brought them into the living room, and then to their bedroom, a cot set up for Brao. Chel was astounded at the size of the house. She had no idea Americans lived so lavishly. DeAndre chuckled, assuring her that by American standards, the house was quite modest.

His mother was cooking a big dinner—meatloaf, mashed potatoes with brown gravy, green beans with tomato sauce, biscuits with honey, and sweet iced tea. With many smiles and words of encouragement, she had her daughter-in-law join her in the kitchen. The two women hit it off, figuring out ways to circumvent the fact neither spoke the other's language. A loving relationship was already forming between the two. With a wink and a broad smile, Jada told her son that he had chosen well.

While Mrs. Washington put the finishing touches on dinner, the rest of them gathered in the chintz living room. Chel had never sat in anything as soft as the sofa and felt almost as if she was being swallowed by it. DeAndre had done a good job of briefing Pastor Willis, and he patiently explained to Chel that he was like Smithy, her pastor in Kon Tum and Pleiku.

When the doorbell rang, Mrs. Washington rushed from the kitchen to answer it. In came a smiling Dashawn. DeAndre was taken completely by surprise. The brothers hugged, with much backslapping and teasing. When they settled down and sat, DeAndre introduced Chel with many good wishes.

"Sorry I can't stay longer, Bro, but I gotta go. I'll be in church tomorrow, and you'll see we both brought a special gift home from Vietnam." No amount of questioning gave DeAndre any more information. His brother just grinned when asked what he meant and said he would find out in the morning at the worship service. With that, he gave his mom a hug, smiled at Chel again, and made his goodbye to Pastor Willis.

In the morning, as the family walked into the little church with Chel holding Mary, they were greeted by Dashawn and a pretty Vietnamese woman who he introduced as his wife Nga Du. She spoke decent English, telling DeAndre she was happy to meet him at last. Like Chel, she had kept her family name rather than take her husband's, as per their culture's tradition.

"Dashawn, you certainly picked a pretty woman to marry. How did the two of you meet?"

"When I came out of the weeds about two months before my DEROS, I became the company exec. The battalion's rear area was in Ah Nhon. One day, I went over to the 67th Evac Hospital in Qui Nhon to check on one of the company's men who had been wounded the day before. As I walked into his ward, I was surprised to see a Vietnamese woman visiting a Vietnamese soldier. As I walked by, she looked up and smiled at me. That, my dear brother, ended it for me—I was already in love," giving his wife a hug.

With that, they all moved into the sanctuary and sat down in Jada's pew. Her face split with pride as the worshipers turned to see her family, grandchildren and all.

After they arrived back at Mrs. Washington's house, the conversation continued. "So, what's the rest of the story?" grinned DeAndre. "And, I don't remember seeing any Vietnamese in US hospitals, except the occasional civilian."

"I'm like you. As I said, I was surprised. From what I understand, some exceptions were made for ARVN who had wounds that the local Vietnamese hospital couldn't handle. From what Nga told me later, the soldier had a severe head injury and was wounded close to the 67th Evac which had neurosurgical capabilities."

With that, Mrs. Washington took over. "Okay, you two. Enough shop talk. Ladies—join me in the kitchen please." Chel put Mary in DeAndre's arms and left Brao playing with toy cars on the floor.

The brothers sat next to each other on the couch, thinking of all the things they needed to cover to bring each other up to date. Pastor Willis listened intently from across the room. When the subject of DeAndre's "capture" and becoming a POW came up, he fibbed to his brother and told him the manufactured story, and how he had been saved by the Montagnards when they ambushed the NVA patrol. "That's how I got to know them better."

"Them?"

"Oh yeah. In Kon Tum, there were a bunch of Jarai villages around, including a section of refugees in the city."

"Jarai?"

With that, DeAndre went on to explain all the different tribes that make up the Montagnards and how he met Smithy and Lee Anne. "So, even though I barely learned the language, I am at least familiar with the Jarai and their customs. They really are different from the Kinh."

"Kinh?"

"Yeah, ethnic Vietnamese. Nga is Kinh. The Montagnards are Vietnamese in nationality only. Few speak Vietnamese and they rebelled against the government—both governments. The men worked with Special Forces and made damned fine fighters. When the south fell, they fought the Communists tooth and nail, but over time, they just ran out of guns, ammo, and men. That's why they're here now. Notice all the Americans working with them are former Green Berets."

"I hadn't noticed because I haven't visited the settlement. But it sounds like you and your wife got down to business fast." Dashawn said with a grin.

"Oh, Mary is our second, and this little boy is named Brao.

"Okay, gentlemen—the dinner bell has rung," announced Mrs. Washington. "Just like when you were kids, wash your hands first."

Pastor Willis added, "Thanks for the anthropology lesson, you two. I'd never heard of Montagnards before, much less the Jarai."

Dinner was as excellent as the brothers expected. Their mother had prepared the dinner just for them—"Larapin" was Dashawn's praise. Dessert was pecan pie, in all its sweet sugary glory. DeAndre leaned over to tell his wife she didn't have to eat all of it, knowing her body was not used to so much sugar. He made excuses to his mother telling her Chel was already full, which was true.

The chatter around the table was slowed by the use of three languages, but the smiles were frequent. Pastor Willis took part a little more, especially when he found out the asylum-seekers were Christian. Jada Washington was just so delighted to see both her boys married off, even if neither married American girls. Her high spirits allowed her to accept some good-natured teasing as she tried to say "Nga" correctly.

"Just remember the word "song" ends with the ng sound. Now begin a word the same way." The laughter as she struggled was not humiliating. It was "Welcome to our world" laughter.

Both new wives vowed to return to their mother-in-law's home to cook a meal from their own cultures. But

DeAndre also noticed a peculiar reserve between Chel and Nga.

Finally, Pastor Willis hinted that it was his bedtime. Jada Washington didn't disagree, so the evening ended pleasantly with friendly chatter among them. But there was a hint that all was not right when DeAndre overheard his sister-in-law use the word "moi" with Dashawn.

While everyone continued to chatter in the driveway, DeAndre asked Nga if her family was from Quy Nhon. "No. My father's home village is Tam Kỳ in Quang Nam Province. But he was in the army, so he moved around a lot. My mother and siblings stayed there while my father changed assignments. Our whole family moved to Saigon when the south began to fall, and my father was in charge of the ARVN in the Saigon area."

DeAndre now knew his brother's wife was the daughter of the hated General Ngô.

Pastor Willis called DeAndre over to his car's driver's window. "There's more people you want to see. Ask your Mom when I might bring Eris Wozniak and his wife over as well as Pastor and Mrs. Smith."

"Oh wow!!! Yeah, let me do that right now!!"

After a quick and excited conversation with his mother, DeAndre told the preacher that the evening after next would be good.

And, two evenings later, Mrs. Washington learned more about her older son's life. "Mom, please meet the man I worked with on our advisor team. This is Eric Wozniak and his wife Heather." He made similar introductions to Chel. "Please, please, all of you be seated," smiled Jada.

The five had started to get to know each other when the doorbell rang.

DeAndre escorted Smithy and Lee Anne in. Her husband had kept the meeting of the Smiths secret from Chel. The surprise was total. Reverting to her native tongue, she shrieked her welcome to the two people who had encouraged DeAndre and her. When the introductions got to Pastor Willis, she took over saying, "This preacher who married us."

To her mother-in-law she said, "These people were . . ." she struggled to find the right word. After a word in Jarai with Smithy, Chel continued, "They encourage DeAndre and me." "Wonderful!!"

Turning to Smithy, Jada said, "I feel your encouragement as a sign from God that He has blessed this union." Eric Wozniak's reaction was to say, "I've certainly heard about you many times, Pastor. It is a true pleasure.

Another triumph of Southern cooking. Fried chicken, along with mashed potatoes and gravy, collard greens, and cornbread more than delighted her guests. Peach cobbler with ice cream completed the feast.

"So, how's retirement?" DeAndre asked his former NCO. "Well, sir, I like it. Nice to be around my wife all the time."

"Okay—but knock off the 'sir' stuff," grinned DeAndre. "We're not in the Army anymore."

"Old habits die hard—Sir."

Chel was so happy to see Smithy and Lee Anne again, and part of her delight was to be able to speak Jarai. Again, Pastor Willis was exposed to another world. De-

Andre was also to find out that besides Smithy and Lee Anne, Wozniak knew the real story of his capture, having learned it from Smithy.

As the happy chatter continued, the doorbell rang. There stood Dashawn and Nga. More introductions. More pleasant chatter. "Sorry we can't stay longer, but we have another engagement. We invite you all to join us for dinner at our house soon. Nga will cook Vietnamese food, and we'll all get to know each other even better."

The invitation was accepted by everyone.

Thursday arrived. DeAndre had borrowed a car from one of the former Special Forces men and took his wife, mother, and pastor to his brother's house. The Wozniaks picked up the Smiths and followed in their car.

The house was nothing fancy; just a simple red brick three-bedroom two bath ranch style place with a garage. DeAndre noticed the perfectly manicured lawn. His brother always had been the persnickety one.

They seemed to be the first guests, and they decided to go in before the others arrived. A ring of the doorbell and the door was opened by Dashawn. A smiling Nga, clad in a beautiful Vietnamese ao dai stood in greeting at the door. Her outfit contrasted delightfully with Chel's Jarai garb.

As they came in and the front door closed, Nga said she would like to introduce her father. Speaking over her shoulder, she said simply, "Ba oi." After some footsteps, he turned the corner.

It was the despised General Ngô himself. He instantly recognized DeAndre. He said something to his daughter in Vietnamese and at the same time, DeAndre told her introductions were not necessary—they knew each other. Eric Wozniak gasped in surprise and could be heard mumbling something low that sounded vaguely like the word "asshole."

Jada, Pastor Willis, Dashawn, Smithy, Lee Anne, and Nga all saw it at the same time: they not only knew each other but obviously there was a lot of hate between the two men. The general pointed at Chel and spoke sharply to his daughter. It did not take a translator to understand he'd said something about her letting "moi" into her house. He spit the word out. DeAndre turned on his heel, telling his mother, brother, the Smiths, and his pastor they were leaving.

By this time, Smithy had figured out the father's identity, but the others were confused as they hesitatingly turned towards the door. DeAndre walked past them to open the door, but Wozniak beat him to it. As they were getting into their cars, the general came to the open door and yelled "Fulro!"

Smithy had better control of his temper than DeAndre, so he took on the task of explaining who General Ngô was—and why he was so despised. As he listened, DeAndre wondered to himself how he would be able to be a true brother to DaShawn. If they lost that connection, it would be hard for their mother. As Pastor Willis listened, he knew there was a lot of family enmity that would need to be healed.

The preacher called DeAndre the next day. "This is going to put a huge barrier between you and Dashawn. For your mother's sake, I hope you two can figure out some sort of truce or even peace. I'll be more than happy to be in the middle between you."

DeAndre wasn't surprised at Pastor Willis' offer. He not only took him up on it but told the minister how much he trusted him.

Dashawn got the exact opposite. He was trapped in the same house with his outraged father-in-law and had to listen to his tirade towards Nga. At one point, he'd had enough and strode towards the general, but Nga took his arm, saying, "No, in my family, the bride's father still outranks a husband. You'll just add fuel to the fire if you try to rescue me."

Dashawn slowly walked away but made sure the general saw the fire—even challenge –in his eyes. It was an uneasy house.

Chel sensed the tenseness in her husband—and she knew why. She wanted her husband and his brother to make amends, letting DeAndre, Dashawn, and their mother be a family again. She asked Smithy and Lee Anne for help. The next day, the church pastor and the missionary were at the door. Pastor Willis spent an hour or so with a shattered and frightened Jada Washington, telling her that he and Smithy were going to try to negotiate peace between the brothers. Then he joined Smithy and DeAndre in the living room.

Pastor Willis began, "Obviously, I've never been in Vietnam, nor have I been in combat, but I know you two were happy brothers who grew up in a loving and secure family." Smithy went on, "And I don't know the Washington family at all, but I am familiar with war and Vietnam."

DeAndre broke the awkward silence that followed. He began, "You two are the most trusted people in my life. You're both true men of God and I have a question for you. It's not about General Ngô per se, but it is part of my dilemma." The two ministers wisely let the silence go on as DeAndre gathered his thoughts.

"Coming home to my mother has been one of the best moments in my life. That doesn't happen to everybody. One time, after a firefight, I was checking the enemy dead. I came across a photo I found on the body of a Vietnamese lieutenant we had shot. It showed him in a dress uniform standing behind two people who were obviously his parents. I realized we were a lot alike. We both had answered the call of our governments to fight. In his case, it was for his country to be free. In my case, it was to protect America from Communism. The photo didn't show any wife, but maybe he had a serious girlfriend he wanted to marry when the war was over. The point is, we were more alike than different. We'd both been blessed with loving parents. We were both educated enough to be officers. What I'm saying is that, as I looked at that photo, I knew God had protected me and I knew God is good.

"But then there is the obvious difference. The Trung uy was dead and I was still alive. Pastor Willis—Smithy—if there is a loving God, why are evil and war allowed to exist?"

Softly—gently—Pastor Willis said, "DeAndre—did God cause the war?

"No, but He allowed it."

"How far along in an argument would God go to allow differences between people before He stepped in to stop it?"

That stumped DeAndre a little, and the two pastors let him chew awhile. Smithy chimed in, "When you were in high school, did you ever have a strong difference of opinion with somebody?"

"Yeah—of course."

"About what?"

"Oh hell—all kinds of things. Who was the best college quarterback? Would all the civil rights protests do any good? Was one guy better than another guy as president of the Key Club? Which girl had the biggest boobs? You know—stuff like that. Nothing important."

"Was it important to you and the others then?"

"Yeah," chuckled DeAndre. "We'd really scream at each other. I remember one time a best friend of mine took a swing at me."

"Did you swing back?"

"Of course . . ." DeAndre trailed off. He got it. A fist fight and war were just a matter of degree.

"So why did God allow that fight?"

Good question thought DeAndre. Finally, he said it out loud, "I don't know."

Now Pastor Willis took up the conversation again. "Before anybody took a swing, did you have a choice to swing or not?"

"Sure."

"Before a war starts, do countries have a choice? For

that matter, did you have a choice as to whether you fought?"

"Yeah, but . . ." began DeAndre, but trailed off as he tried to digest the lesson he was learning.

"Humans are born with choices. True, we are often heavily influenced by our culture, but we still have choices, whether we want to admit it or not." Looking over to Smithy with a smile, Pastor Willis went on, "We professional Christians call that free will."

Smithy went on saying, "You had a choice, DeAndre. You didn't have to go to war. Neither did the Trung uy. Oh, don't get me wrong—either one or both of you would have gone to prison if you chose not to, but you still had a choice."

His childhood pastor finished up. "DeAndre—Jesus had choices too. Remember, he could have waved his hand, and a legion of angels would come down and annihilate the high priest, Pilate and all the other officials. But his choice was to go to his very painful, agonizing death. God loves us and wants us to love Him. If he took away choice, we would be mere robots when we loved him—we would have no choice."

"Okay—I think I understand. General Ngô had and still has a choice, and he has chosen to hate the Jarai. If God stopped that, He would be stripping us of our ability to choose."

Smithy lightened the mood by saying, "Hey Pastor Willis—we must be pretty good teachers!"

Now the conversation could move on to its original

reason and try to patch up the relationship between the two brothers. Now DeAndre realized he had choices.

The conversation went on for two hours, with everyone trying to contribute something worthwhile towards solving the problem. DeAndre wanted to be reconnected to his brother. It was Pastor Willis who suggested the two brothers get together in a neutral place like a coffee shop. Their childhood pastor would be there too, not so much to try to fix the problem, but to keep feathers from getting ruffled.

The good minister worked fast. The very next day, he called Dashawn and DeAndre. They would meet at Union Coffee in downtown Greensboro the following Wednesday morning. They both could tell their wives where they were going, but the information was not to be passed on to Nga's father or DeAndre's mother. The three took turns praying for success.

Wednesday soon came, and at 10, both brothers found Pastor Willis in a cozy booth, enjoying his cup of cafe mocha. They slid into the booth, sitting next to each other, but soon the preacher suggested they get some coffee too.

"Okay, guys. I'm not here to referee a fight. If the most important thing to you is to win this battle, then I'm outta here." This kind of uncharacteristic blunt talk from their childhood clergyman got the brother's attention. "You guys know how to solve this, not me."

The doubtful look on DeAndre's face got this in reply. "You may not even know what that answer is right now. You're both so angry you can't see much except a fight. Be thinking right now—what does your brother want to hear? Have you thought about it? Do it now. In a few moments I will ask you, DeAndre, what Dashawn wants to hear right now. Or maybe I'll ask Dashawn first. Get to it—think what the other wants to hear from you."

With that, he excused himself to get another cup of coffee. Bringing back three pastries along with his coffee, he sat watching his charges for a few minutes. "Okay, DeAndre. What does Dashawn want to hear right now?"

DeAndre had a slightly panicked look on his face. He'd anticipated a meeting to figure out how to get General Ngô out of the picture, not looking into his own heart. Hesitatingly, he started, "Dashawn, I think you'd like to hear how you and I can be close again." Hedging his bets, he added another idea, "I'm not sure that's what you want. Maybe you may want ideas in calming down your father-in-law," bringing a faint smile to his brother's face.

"I've never seen him like that before—not even close. He's been very nice to me, doesn't seem to mind that his daughter married a black man." Turning to Pastor Willis, he continued, "Do you want me to give my answer now, or can I ask him some questions?"

"Oh, I'm not into formulas. As long as the questions are not hostile and they are intended to actually solve the problem, let's go with your questions."

Dashawn took a moment or two to collect his thoughts. "I swear, DeAndre—neither Nga or I had any idea he was going to go off the deep end like that. Nga was as surprised as I was. I mean, she understands the word moi, and may have used it once in a while, but she honestly doesn't hate the minority people."

He paused and took a deep breath, letting it out slowly. "It was obvious just as soon as General Ngô saw you and Chel that he was crazy angry. And you seemed to hate him too. But neither of us have any idea why you two hate each other so much. And yes, I heard Eric say the word asshole under his breath."

DeAndre immediately saw that he had just automatically assumed everyone would know of Ngô's hatred of the Dega and would therefore understand the standoff. "Pastor Willis—help me out with this. I could answer my brother, but I don't know if I could control my invective enough to be objective. What would you think of my asking Smithy to answer that question?"

"That's an easy one to answer, DeAndre." With that, he left the booth, walked to the front door, let out a wolf whistle and signaled someone to come inside. Smithy willingly told the others of his own experiences and what he had heard about General Ngô, including his direct supervision of killings by ARVN troops, especially by the Rangers under Hoang. Smithy went on to say that according to one spy, Ngô had told his commanders that killing FULRO was more important than killing Communists. Twenty minutes later, he ended the session with the story of Ngô using helicopter gunships to fire on Montagnard civilians.

Pastor Willis and Dashawn were both truly taken aback by a respected minister of God painting a man as consummate evil. DeAndre took the chance to reach over and rest his hand on his brother's shoulder, who looked genuinely shocked at the allegations made against his father-in-law.

Turning to DeAndre, his brother said, "My brother, I love Nga very much. Very much indeed. I don't know if she knows any of this. If she doesn't, that would explain why she looked so shocked by her father's words the other day. He came to North Carolina well after we did simply because, as so-called boat people, he was put in a refugee camp. When Nga found out he was at Fort Chaffee, she drove there and worked out an arrangement for him to live with us. He's been here for about four or five months."

The tension broken, the brothers decided to meet alone the next day. Besides trying to solve the problem, the two began to share their experiences, soon laughing or agreeing or identifying with the other's personal lives in Vietnam. Both had told their wives where they were and about their efforts to reach a rapprochement. The next day, their wives joined them, giving them a chance to get to know their new sister-in-law.

Unbeknownst to Nga, Ngô's former aide-de-camp in the old army, had followed her to the cafe. Being sure his face was hidden; he went in and saw the four of them laughing and enjoying themselves and dutifully reported that to his boss.

The general was sugary sweet when Nga came home, gently asking if she'd had a good time. She wasn't suspicious, but rather mildly curious.

Trying not to set a pattern, the four had decided not to meet the next day, but rather in two days hence. When Nga left the house to drive herself to the cafe, Ngô called his aide who picked up General Ngô and drove him to the cafe.

It was not a subtle entry. He banged open the door and looked around until he spotted his daughter in the booth with the hated Chel. Loudly striding over to them, he grabbed Chel's hand and yanked—hard—and she tumbled to the floor. Ngô screamed Vietnamese obscenities as he stood over her and kicked her in the ribs. DeAndre leaped from his seat, landing atop the general's shoulders. The two yelled at each other as they landed on the floor.

Fists met jaws, fingers probed throats, and knees found groins. No amount of pleading from others made the slightest dent in the hatred the two men had for each other. Nga went to Chel to see if she was badly hurt. She wasn't. Within a few minutes, the police arrived and broke up the combatants. They were handcuffed and put into the back seat of separate patrol cars.

DeAndre was to find out later how Pastor Willis found out he was behind bars, but the preacher was at the jail within a few minutes. It seems Ben Cooper of Save the Montagnard People was seated with his wife on the other side of the room. He didn't know who Ngô was, but he sure knew who DeAndre was. Ben made a quick phone call to the pastor, picked him up, and both went to the jail.

A detective was talking to DeAndre. After waiting for the officer to finish, Pastor Willis sat outside the cell. Ben had offered that he spoke some Vietnamese and would be happy to assist the police with questioning Ngô.

Police questioning was primarily with DeAndre because Ngô had no knowledge whatsoever of how the criminal justice system worked in America. The first thing he told Ben was to find out was how much the detective and the head policeman wanted as a bribe. It took a while for the former general to understand he had no authority in the US at all. None.

"Pastor, I really, really don't want anything to happen to General Ngô. His racism is so over the top and he is so out of his element that the only thing he could do is to leave here. If he does that, I hope he will let his daughter Nga stay here."

The detective listened to that conversation and interjected. "I'm just trying to figure out how this fight started. From what I understand, the Vietnamese guy just barged into the cafe and went for Mr. Washington's wife right off the bat. What was that all about?"

"DeAndre, I think I've heard your story enough times to be able to tell it. Detective Franklin can ask you for clarification if need be. That okay, Detective?"

"That's fine with me, Pastor. Mr. Washington—chime in when you need to."

"Okay. There's a lot of background in the story, so I hope you're not in a hurry." With that, Pastor Willis explained why DeAndre was in the mountains of Vietnam and how he had gotten very tired of Thiếu tá Hoang's brutality towards the Jarai (and gave Smithy's name as a backup). When he became the senior advisor of the region, General Ngô showed his violent prejudice and sic'd a helicopter gunship on unarmed Jarai civilians. He told of DeAndre meeting Chel and the support they got from Smithy and Lee Anne. Finally, Pastor Willis told of De-

Andre "going native" as a fighter with Fulro. The whole story took the best part of 45 minutes, but Detective Franklin thanked the good reverend profusely.

"What's next, Detective?"

"Hang on. I want to talk to our head of detectives first."

When he returned, saying that he had not only met with his boss, but the chief of police as well. "I want to talk to you, plus Ben, the pastor, and any other relevant person to see what can be done. I just talked to the chief, and nobody is crazy about taking you or Ngô to trial. The general has no idea what's going on."

"I suggest you include both my brother and his wife Nga in the conversation."

"Good idea. We can do that today or we can do it tomorrow, but if we wait, I will have to release him. I can't hold him in jail with no charges."

An hour and a half later, the big, polished table in the police department's conference room was surrounded by all the principals, except Ngô himself. Nga thought it best for the group to come up with a proposed solution and she relay it to him."

"Okay. But listen, everybody. The best we can do by ourselves is to propose a plan and present it to the municipal judge. He will have to make the final decision, and I'm sure he will want to know if there was damage to the cafe, if anybody was hurt, and other matters of the law."

Dashawn and Nga had talked it over before arriving at the police building and offered the first idea.

"Let's see if the judge, DeAndre, and the cafe owner

have any criminal charges they want to press. If not, then we propose he be told to leave North Carolina."

Nga interjected, "My father has many family members in San Jose, California, which has a fairly large population of refugee Vietnamese. I will call ahead to his brother and other family members to see if they are okay with the idea of him living there. I see no reason why anyone would object. My father won't have to live with anybody. He has enough money to live on his own—he'd just need a little help in finding a place. I will take him to San Jose myself."

Smiles erupted around the table. "Great solution, Nga." With everyone agreeing, Detective Franklin went to see the judge, who immediately convened a hearing in the courtroom. Pastor Willis drove over to the cafe and asked the owner to come to the hearing, explaining Nga's idea to him in the car.

"The court will come to order," said the judge. The conversations quickly died down, and Judge Wilshire took his seat. "Detective Franklin. I understand you worked an assault and battery case earlier today and the principals involved have some suggestions as to how the case can be disposed. I ask everyone to keep in mind that this is not a trial. Any testimony will not be sworn and no records taken. We are merely trying to find a workable solution that everyone can agree on."

"Your honor—I got a call from uniformed officers about a fight at the Union Cafe on 2nd Street. By the time I got there, two men had been separated, cuffed, and placed in cruisers." Indicating the others, he continued, "I talked to Mr. and Mrs. Deshawn Washington, Mrs. Chel

Washington, wife of one of the arrestees, Pastor Willis of the Resurrection Baptist Church, and Ben Cooper, former Green Beret and now with Save the Montagnard People. The background makes for a long story. Since these folks know much more about the circumstances, I request your permission to ask them questions, so you get a more succinct answer."

"Of course, Detective. Proceed."

"First of all, your honor, notice the differences between the two Mrs. Washingtons. One is obviously Vietnamese and the other has fewer Asian features and darker skin. They are dressed quite differently. One is an ethnic Vietnamese and the other is from the mountain area of Vietnam and is ethnically what is called a Montagnard. Notice too that Mr. Washington is black. He is the brother of one of the arrestees, and his wife is the daughter of the other arrestee. There is a longstanding history of discrimination by ethnic Vietnamese against the Montagnards.

"That discrimination has become more pronounced once it was discovered the mountains are great places to grow coffee and tea, but the underlying basis for the discrimination is simple racism. Your honor, I invite you to think of the native Americans in this country and how their land was taken by the white man over the course of centuries, usually by violence and killings."

With that, he asked the others questions about their role in the current situation. After twenty minutes, he ended it by saying, "So, your honor, General Ngô was very upset that his daughter would actually invite a 'moi' into her home. One can understand the virulence of his anger when you consider he had directed the killing of Montagnards by his troops, and on occasion, had killed them himself."

"That's quite a story, Detective." With that, the judge asked a few clarifying questions of the people involved. "Folks, I'm going to come down from the bench and sit with you all." Standing, he removed his robe, then asked his bailiff if the conference room was empty. It was and they all walked down the hall.

"I understand that Mr. DeAndre Washington, who was arrested, is not interested in having the general go to prison. That's quite magnanimous."

"He doesn't want to see his brother's father-in-law in prison. This is a family matter, your honor," said Pastor Willis. "If he goes to prison, that will almost certainly alienate the brothers, which of course would be very hard on their mother, who is a widow."

The judge took a few moments to gather his thoughts. "Okay. I can't make this a ruling of the court. I'm simply one of the group trying to solve this problem. Frankly, I think it's a great idea to send him off to San Jose. He'll be surrounded by family and in a culture that will be much easier to adapt to than here. Give me a few hours, but I will bring together the courtroom people needed for me to make a ruling that both men be released stating an unwillingness to prosecute. I will call the court to order at 6."

The following day, General Ngô was on an airplane, along with his aide-de-camp, bound for San Jose. Nga was not with him as he didn't want his "moi loving" daughter with him right then. DeAndre was released from jail, enthusiastically greeted by Chel at the front door of the jail. After many "I love you" and "I missed you," they got in Pastor Willis' car for the drive to Mrs. Washington's home. There they were joined by Dashawn and Nga, and along with the pastor, enjoyed a feast prepared by a very happy mother.

"I was really worried my sons would fight with each other over this thing," Mrs. Washington said, refusing to give "this thing" a name.

"And we owe it to Nga," said DeAndre. "It was her idea to send her father to San Jose." Turning to his sister-in-law, he said, "And I hope you don't lose your relationship with your father."

"I think we are okay. He wouldn't dream that his own daughter would banish him. I'll go out to San Jose in a couple of weeks or so just to check on him."

DeAndre stood, then raised his water glass, saying, "To Nga." Everyone echoed, "To Nga" as she blushed and smiled. The rest of the evening was very pleasant as Mrs. Washington got to know her two new daughters-in-law.

CHAPTER TWENTY-SIX

Now began a busy time for Chel as she had classes to attend to learn English and American culture. Both brothers got jobs, Dashawn with Cone Health as a patient advocate, and DeAndre with Dr. Raleigh Bailey, founder of the Center for New North Carolinians at the University of North Carolina at Greensboro, as a traveling researcher.

DeAndre bought a car. Chel began to see American houses, and she wanted to live in one rather than a longhouse. She dressed the two children in typical American style and proudly sent Brao off to school. The brothers spent many Saturdays at each other's homes, watching sports, but Sunday morning was reserved for worship with Mrs. Washington, followed by one of her signature feasts.

Both brothers had jobs where they were in contact with new refugees, although the number grew smaller as the number of asylum seekers declined. The mountain people were either unable to leave Vietnam, or the government of Cambodia was trying to force the Montagnards there back to Vietnam.

By the year 2000, the few people who could escape the Communist police did so by going through Laos, then to

Thailand, and finally into the U.S. DeAndre made it his business to meet these people to glean what information he could about current conditions. What he heard horrified him. The government effectively tried to close the Christian churches. The tribes reacted by staging large, peaceful protests. The government reacted harshly by bringing in hundreds of police and soldiers.

Over the following months, hundreds of highlanders were arrested. Torture was sometimes used to get confessions and find other protesters who had gone into hiding. Some religious leaders were sentenced to prison, some for as long as twelve years. Many church buildings were torched. Many of the protesters who fled to Cambodia were pursued by Vietnamese forces, with many Highlanders forced back to Vietnam.

DeAndre met many of the escapees, some of whom needed medical attention. He also met Kok Ksor, a Jarai living in the U.S. and the founder of the Montagnard Foundation Inc. He was amazed to hear an American black man speak fluent Jarai but then put two and two together and realized this was the man who had once been a captain in the American army and a FULRO fighter. Though Kok Ksor had also been in FULRO, he said that he no longer advocated violence but instead sought to help his people back in Vietnam through education, health care, and legal means.

DeAndre had already heard many stories of the brutal repression of all the Dega, but Kok Ksor had even more stories. The two quickly became friends, DeAndre even inviting him to his home where Chel made a proper Jarai meal.

As he profoundly praised Chel's cooking, he spoke to both her and DeAndre about the Montagnard Foundation. "I thought it was bad when I left the mountains, but in talking to the people coming in now, it has really gotten bad." Reaching for a second helping, he went on, "I have told you that MFI doesn't advocate violence—and that's true. But I also do a little quiet recruiting for FULRO."

"I thought FULRO was dead," said DeAndre. "Didn't it go away a few years ago?"

"Yes, it did. But, shall we say, it is being resurrected. We have to be super quiet about it. If the US government were to find out we are smuggling weapons into Vietnam, it would probably stop all assistance to MFI. I have to keep FULRO and MFI separate."

"Makes sense. Has FULRO done any operations yet?"

"Come on, Captain! I would be violating all kinds of operational security if I told you that." Kok Ksor grinned. "But, yes—very, very small operations, but they hope to launch bigger operations as our fighters gather experience. Plus, they have to operate outside of their villages as we don't want to put civilians in any more danger than they already are."

Chel watched her husband with intensity. He was digging for details, and she could almost see his military mind twirling. When DeAndre invited Kok Ksor back for another dinner, she became even more watchful.

The door had barely closed when he turned to see

his wife, hands on hips, and a big scowl. "What are you thinking, DeAndre?" she asked.

As always, Chel was a step ahead of her husband. "When Kok Ksor comes over the next time, I will be in on the conversation." DeAndre noticed that he had no choice in the matter—Chel would be talking to the FULRO advocate too.

Two weeks later, Kok Ksor returned for both Chel's cooking and to talk about FULRO. Food on the table, and after a few minutes of chit-chat, Chel asked, "Have you been able to recruit any FULRO fighters here in North Carolina?"

The directness of the question unsettled DeAndre, but he didn't interrupt. He saw that his guest was caught off guard too, but after a few moments to gather his thoughts, he answered, "Yes, we have, but not many. Most of them had been fighters, but few were leaders and nobody who led larger groups. Again, I have to ask you both to not tell anybody—not even relatives. If General Ngô ever heard about this, he would probably organize hit squads."

"Not to worry. I won't put my brother into that uncomfortable position." DeAndre looked at his wife, a thin smile on his face. "But, if I'm correct, what my wife really wants to know is if you are trying to recruit me." A quick smile from Chel told him he was right in that suspicion.

"Oh, wow! No, I hadn't thought of that. After all, you have two kids and most likely Chel would stay with them. It's too much to ask anybody to leave his family to go fight again."

"Or take them with us," said Chel.

Kok Ksor sensed he had unwittingly wound up in the middle of a truly heavy-duty conversation, and it made him very uncomfortable. He soon made his excuses and left the house.

"Are you serious? You'd take Brao and Mary back to Cambodia in a very, very hostile situation?"

"DeAndre. You know how very much I love you. But I also know that you sometimes have to lean on me for support. That's what husbands and wives do—love and support each other."

With that, DeAndre walked over to his wife and gave her a passionate kiss. It was returned enthusiastically. Hoarsely she whispered, "Let's talk about this some more after we take the edge off."

When they awoke, Brao came home from school and DeAndre left to pick up Mary at daycare. While making the evening meal together, she said she wanted to ask her lady friends if they had any information of what was happening back in Cambodia and Vietnam. Brao watched from the other room as his parents laughed and touched each other.

Three months later, Kok Ksor stood at the gate watching a jet taxi towards the runway. Aboard were all four members of the Washington family. It was bound for Dallas Ft. Worth Airport, then another plane to Taiwan with a final switch to a plane to Phnom Phen, Cambodia.

It took a few days for Chel to figure out how to get her family from Phnom Phen to the Highlands, and even longer to negotiate the obstacles to their journey to Chel's home village. Her parents were delighted to see their grandchildren.

But both Chel and DeAndre knew they would be safer if they were in Cambodia. They knew where they could find a FULRO camp. To the great enjoyment of the villagers, they watched as Mary and Brao grew up—and DeAndre and Chel continued their love affair.

EPILOGUE

According to reliable news sources such as the Voice of America, The Diplomat, and the BBC, there was a coordinated attack in June 2003 against government and police buildings in Dak Lak Province. A total of nine officials were killed, including four police officers. One hundred people were put on trial in January 2004. No list of arrestees has been published.

ACKNOWLEDGMENTS

As always, I have wonderful support from my family, especially my wife Cindy, son Keith Young, daughter Trang Phan and daughter Ai Nhan Ngo.

There were several people who served as beta readers, including my son Keith, old friends Jeff Wilhoit and Jim Barnes, my brother-in-law Mark Mason and good friend Helen Spalding. I thank Dr. William Chickering, author of "A War of Their Own" and Nghieng Nay of the Montagnard Dega Association for their help with Jarai names

ABOUT THE AUTHOR & THIS NOVEL

My life is centered around Vietnam. I served in the infantry there for two year-long tours and have a Purple Heart to show for it. My son Keith was born of my first marriage in 1967 during my first tour. I met my wife Cindy there in 1969 while she was an Army nurse. We returned in 2005-06 to live in the city of Hue, teaching at the university. After returning home, we made many return trips. In 2008 and again in 2011, Cindy and I brought two former students to the United States to earn their master's degrees. Today, both of them have Ph.Ds and both became very close to us. Trang Phan was adopted as an adult in 2015 and Ai Nhan Ngo in 2017.

While teaching at the University of Hue, I had one class composed of students from minority tribes in the mountains of Central Vietnam. It was my favorite group of students. English was their third language, Vietnamese being their second and their tribal language being first. The regular faculty did not want to teach them. I also had a graduate class and would sometimes meet them for Saturday morning coffee. Many of them were teachers in the area around Buon Ma Thuot and many of their students were from minority tribes. I was deeply surprised when the grad students said things like, "I don't know why we even try to educate them—they're too stupid to learn." I heard the word "moi" bandied about and found the word meant "savage."

The average Vietnamese despises the minorities simply because they didn't know them. There are no tribal members living next door to Vietnamese families. In my "American Culture" class, when shown photos of black or Asian Americans, they never identified the blacks or Asians as Americans—only whites were Americans. Theirs is very monocultural society

I was drawn to one minority student in particular. His name was Ngok, which is usually spelled Ngoc. He also wore a cross, indicating he was Christian. I once asked him why he spelled his name with a k. When I saw that he was slow to answer, I asked, "Do you do it to piss off the Kinh? (ethnic Vietnamese)" He smiled quickly and merely answered yes. Today, he is a teacher near the city of Pleiku in the mountains, and I occasionally swap emails with him.

BIBLIOGRAPHY

We Have Eaten the Forest by Georges Condominas (English translation), originally written in French and published in 1957. The English version was published in 1977. The author was a Frenchman born in Hanoi. He was a trained anthropologist, earning his doctorate at The Sorbonne.

Window on a War by Gerald Hickey. Published in 2002 by Texas Tech University, most of the book concerns the Montagnard people during the American war in Vietnam. Hickey has an earned PhD in anthropology.

To Vietnam with Love: The Story of Charlie and EG Long. Published 1995 by Christian Publications Inc., it is the autobiography of the missionary who is the basis for "Smithy." Charlie and his wife EG lived among the Jarai, learned the language and translated both the New Testament and a hymnbook into Jarai. They had their scrapes with Communist forces. Upon their return to the US, he ministered to the Jarai refugees who settled in western North Carolina. I created Smithy and Lee Ann's names because I have inserted some activities that the real people never did.

The Montagnards — Culture Profile - originally a pamphlet created by Dr. Raleigh Bailey, founding director of the Center for New North Carolinians at the University of North Carolina at Greensboro. https://cnnc.uncg.edu/wp-content/uploads/2015/06/COR-Montagnard-Profile.pdf

A Bright Shining Lie: John Paul Vann and America in Vietnam—by Neil Sheehan. On pages 754 through 760, there is a view of the battle of Kon Tum as witnessed by Vann. Published by Vintage Books in 1988, this is one of the seminal books about the war.

Vietnam Order of Battle—by Shelby Stanton. This is one of the very best reference books of the conflict. I used it in this book to establish where American units were based and when they departed Vietnam.

Tiger Men—by Australian Barry Petersen who had been an Australian Army captain advising Montagnard fighters. Published in 1988 in Melbourne. The story of the attack on a Kinh village that was in the open was based on this book.

In Retrospect: The Tragedy and Lessons of Vietnam by former Secretary of Defense Robert S. McNamara and Brian VanDeMark, Vintage Books, 1996. Way back in 1963, McNamara had said it was a Vietnamese war. President Nixon revived the idea in the 1970s and called it "Vietnamization.

How to Stay Alive in Vietnam: What It Takes to Survive in This Different Kind of War Robert B. Rigg, 1966.

USEFUL LINKS

Save the Montagnard People - https://www.montagnards.org/

Nate Thayer's FULRO's Jungle Christians - https://natethayer.wordpress.com/2013/10/28/vietnam-era-renegade-army-discovered-lighting-the-darkness-fulros-jungle-christians/

Veterans of Foreign Wars - Montagnards Find Home in North Carolina - https://www.vfw.org/media-and-events/latest-releases/archives/2019/4/montagnards-find-home-in-north-carolina

Los Angeles Times - Celebrate First Thanksgiving: Montagnard Refugees Find New Life in U.S. This includes the story of Rahma Dock. https://www.latimes.com/archives/la-xpm-1986-11-30-mn-481-story.html

Wikipedia - FULRO - https://en.wikipedia.org/wiki/United_Front_for_the_Liberation_of_Oppressed_Races#cite_note-O'Dowd2007_2-40

BBC - Vietnam War: The pastor who survived 17 years

in forgotten jungle army - https://www.bbc.com/news/world-asia-63374482

CIA - Vietnamese Fear a Tribal Uprising (dated April 1966)

https://www.cia.gov/readingroom/document/cia-rdp73-00475r000201600001-8

BBC - Dega Attack on Police Station - https://www.bbc.com/news/world-asia-67995372 This BBC article was written just before the trial of tribesmen accused of killing nine Kinh police officers in June 2023.

United Nations High Commission on Refugees: Repression Of Montagnards: Conflicts over Land and Religion in Vietnam's Central Highlands

Dr. Patricia Smith - A real person who ran a charity hospital in Kon Tum from 1959 to 1975. She left in April 1975, but two of her colleagues, Dr. George Christian, from Massachusetts, and a New Zealander, Dr. Edric Baker decided to stay behind. https://www.nytimes.com/1975/04/04/archives/us-woman-doctor-leaves-vietnam-after-16-years.html

www.ingramcontent.com/pod-product-compliance
Lightning Source LLC
LaVergne TN
LVHW010639110826
845149LV00014B/2884

9781961505582